Introduction to Deep Learning Using R

A Step-by-Step Guide to
Learning and Implementing
Deep Learning Models Using R

Taweh Beysolow II

Apress®

Introduction to Deep Learning Using R

Taweh Beysolow II
San Francisco, California, USA

ISBN-13 (pbk): 978-1-4842-2733-6　　　　ISBN-13 (electronic): 978-1-4842-2734-3
DOI 10.1007/978-1-4842-2734-3

Library of Congress Control Number: 2017947908

Copyright © 2017 by Taweh Beysolow II

This work is subject to copyright. All rights are reserved by the Publisher, whether the whole or part of the material is concerned, specifically the rights of translation, reprinting, reuse of illustrations, recitation, broadcasting, reproduction on microfilms or in any other physical way, and transmission or information storage and retrieval, electronic adaptation, computer software, or by similar or dissimilar methodology now known or hereafter developed.

Trademarked names, logos, and images may appear in this book. Rather than use a trademark symbol with every occurrence of a trademarked name, logo, or image we use the names, logos, and images only in an editorial fashion and to the benefit of the trademark owner, with no intention of infringement of the trademark.

The use in this publication of trade names, trademarks, service marks, and similar terms, even if they are not identified as such, is not to be taken as an expression of opinion as to whether or not they are subject to proprietary rights.

While the advice and information in this book are believed to be true and accurate at the date of publication, neither the authors nor the editors nor the publisher can accept any legal responsibility for any errors or omissions that may be made. The publisher makes no warranty, express or implied, with respect to the material contained herein.

Cover image designed by Freepik

    Managing Director: Welmoed Spahr
    Editorial Director: Todd Green
    Acquisitions Editor: Celestin Suresh John
    Development Editor: Laura Berendson
    Technical Reviewer: Somil Asthana
    Coordinating Editor: Sanchita Mandal
    Copy Editor: Corbin Collins
    Compositor: SPi Global
    Indexer: SPi Global
    Artist: SPi Global

Distributed to the book trade worldwide by Springer Science+Business Media New York, 233 Spring Street, 6th Floor, New York, NY 10013. Phone 1-800-SPRINGER, fax (201) 348-4505, e-mail orders-ny@springer-sbm.com, or visit www.springeronline.com. Apress Media, LLC is a California LLC and the sole member (owner) is Springer Science + Business Media Finance Inc (SSBM Finance Inc). SSBM Finance Inc is a **Delaware** corporation.

For information on translations, please e-mail rights@apress.com, or visit http://www.apress.com/rights-permissions.

Apress titles may be purchased in bulk for academic, corporate, or promotional use. eBook versions and licenses are also available for most titles. For more information, reference our Print and eBook Bulk Sales web page at http://www.apress.com/bulk-sales.

Any source code or other supplementary material referenced by the author in this book is available to readers on GitHub via the book's product page, located at the following link:

https://github.com/TawehBeysolowII/AnIntroductionToDeepLearning.

For more detailed information, please visit http://www.apress.com/source-code.

Printed on acid-free paper

Contents at a Glance

About the Author xiii

About the Technical Reviewer xv

Acknowledgments xvii

Introduction xix

■Chapter 1: Introduction to Deep Learning 1

■Chapter 2: Mathematical Review 11

■Chapter 3: A Review of Optimization and Machine Learning 45

■Chapter 4: Single and Multilayer Perceptron Models 89

■Chapter 5: Convolutional Neural Networks (CNNs) 101

■Chapter 6: Recurrent Neural Networks (RNNs) 113

■Chapter 7: Autoencoders, Restricted Boltzmann Machines, and Deep Belief Networks 125

■Chapter 8: Experimental Design and Heuristics 137

■Chapter 9: Hardware and Software Suggestions 167

■Chapter 10: Machine Learning Example Problems 171

■Chapter 11: Deep Learning and Other Example Problems 195

■Chapter 12: Closing Statements 219

Index 221

Contents

About the Author xiii

About the Technical Reviewer xv

Acknowledgments xvii

Introduction xix

■Chapter 1: Introduction to Deep Learning 1

Deep Learning Models 3

Single Layer Perceptron Model (SLP) 3

Multilayer Perceptron Model (MLP) 4

Convolutional Neural Networks (CNNs) 5

Recurrent Neural Networks (RNNs) 5

Restricted Boltzmann Machines (RBMs) 6

Deep Belief Networks (DBNs) 6

Other Topics Discussed 7

Experimental Design 7

Feature Selection 7

Applied Machine Learning and Deep Learning 7

History of Deep Learning 7

Summary 9

■Chapter 2: Mathematical Review 11

Statistical Concepts 11

Probability 11

And vs. Or 12

v

Bayes' Theorem 14
Random Variables 14
Variance 15
Standard Deviation 16
Coefficient of Determination (R Squared) 17
Mean Squared Error (MSE) 17

Linear Algebra 17
Scalars and Vectors 17
Properties of Vectors 18
Axioms 19
Subspaces 20
Matrices 20

Summary 43

Chapter 3: A Review of Optimization and Machine Learning 45
Unconstrained Optimization 45
Local Minimizers 47
Global Minimizers 47
Conditions for Local Minimizers 48

Neighborhoods 49
Interior and Boundary Points 50

Machine Learning Methods: Supervised Learning 50
History of Machine Learning 50
What Is an Algorithm? 51

Regression Models 51
Linear Regression 51

Choosing An Appropriate Learning Rate 55
Newton's Method 60
Levenberg-Marquardt Heuristic 61

What Is Multicollinearity? 62
Testing for Multicollinearity 62
Variance Inflation Factor (VIF) 62
Ridge Regression 62
Least Absolute Shrinkage and Selection Operator (LASSO) 63
Comparing Ridge Regression and LASSO 64
Evaluating Regression Models 64
Receiver Operating Characteristic (ROC) Curve 67
Confusion Matrix 68
Limitations to Logistic Regression 69
Support Vector Machine (SVM) 70
Sub-Gradient Method Applied to SVMs 72
Extensions of Support Vector Machines 73
Limitations Associated with SVMs 73
Machine Learning Methods: Unsupervised Learning 74
K-Means Clustering 74
Assignment Step 74
Update Step 75
Limitations of K-Means Clustering 75
Expectation Maximization (EM) Algorithm 76
Expectation Step 77
Maximization Step 77
Decision Tree Learning 78
Classification Trees 79
Regression Trees 80
Limitations of Decision Trees 81
Ensemble Methods and Other Heuristics 82
Gradient Boosting 82
Gradient Boosting Algorithm 82

- Random Forest 83
- Limitations to Random Forests 83

Bayesian Learning 83
- Naïve Bayes Classifier 84
- Limitations Associated with Bayesian Classifiers 84
- Final Comments on Tuning Machine Learning Algorithms 85

Reinforcement Learning 86

Summary 87

Chapter 4: Single and Multilayer Perceptron Models 89

Single Layer Perceptron (SLP) Model 89
- Training the Perceptron Model 90
- Widrow-Hoff (WH) Algorithm 90
- Limitations of Single Perceptron Models 91
- Summary Statistics 94

Multi-Layer Perceptron (MLP) Model 94
- Converging upon a Global Optimum 95
- Back-propagation Algorithm for MLP Models: 95
- Limitations and Considerations for MLP Models 97
- How Many Hidden Layers to Use and How Many Neurons Are in It 99

Summary 100

Chapter 5: Convolutional Neural Networks (CNNs) 101

Structure and Properties of CNNs 101

Components of CNN Architectures 103
- Convolutional Layer 103
- Pooling Layer 105
- Rectified Linear Units (ReLU) Layer 106
- Fully Connected (FC) Layer 106
- Loss Layer 107

Tuning Parameters 108

Notable CNN Architectures 108

Regularization 111

Summary 112

■**Chapter 6: Recurrent Neural Networks (RNNs)** **113**

Fully Recurrent Networks 113

Training RNNs with Back-Propagation Through Time (BPPT) 114

Elman Neural Networks 115

Neural History Compressor 116

Long Short-Term Memory (LSTM) 116

Traditional LSTM 118

Training LSTMs 118

Structural Damping Within RNNs 119

Tuning Parameter Update Algorithm 119

Practical Example of RNN: Pattern Detection 120

Summary 124

■**Chapter 7: Autoencoders, Restricted Boltzmann Machines, and Deep Belief Networks** **125**

Autoencoders 125

Linear Autoencoders vs. Principal Components Analysis (PCA) 126

Restricted Boltzmann Machines 127

Contrastive Divergence (CD) Learning 129

Momentum Within RBMs 132

Weight Decay 133

Sparsity 133

    No. and Type Hidden Units 133

Deep Belief Networks (DBNs) 134
Fast Learning Algorithm (Hinton and Osindero 2006) 135
    Algorithm Steps 136
Summary 136

Chapter 8: Experimental Design and Heuristics 137
Analysis of Variance (ANOVA) 137
    One-Way ANOVA 137
    Two-Way (Multiple-Way) ANOVA 137
    Mixed-Design ANOVA 138
    Multivariate ANOVA (MANOVA) 138
F-Statistic and F-Distribution 138
    Fisher's Principles 144
Plackett-Burman Designs 146
Space Filling 147
Full Factorial 147
Halton, Faure, and Sobol Sequences 148
A/B Testing 148
    Simple Two-Sample A/B Test 149
    Beta-Binomial Hierarchical Model for A/B Testing 149
Feature/Variable Selection Techniques 151
    Backwards and Forward Selection 151
    Principal Component Analysis (PCA) 152
    Factor Analysis 154
    Limitations of Factor Analysis 155
Handling Categorical Data 155
    Encoding Factor Levels 156
    Categorical Label Problems: Too Numerous Levels 156
    Canonical Correlation Analysis (CCA) 156

Wrappers, Filters, and Embedded (WFE) Algorithms 157

Relief Algorithm 157

Other Local Search Methods 157

Hill Climbing Search Methods 158

Genetic Algorithms (GAs) 158

Simulated Annealing (SA) 159

Ant Colony Optimization (ACO) 159

Variable Neighborhood Search (VNS) 160

Reactive Search Optimization (RSO) 161

Reactive Prohibitions 162

Fixed Tabu Search 163

Reactive Tabu Search (RTS) 164

WalkSAT Algorithm 165

K-Nearest Neighbors (KNN) 165

Summary 166

■Chapter 9: Hardware and Software Suggestions 167

Processing Data with Standard Hardware 167

Solid State Drives and Hard Drive Disks (HDD) 167

Graphics Processing Unit (GPU) 168

Central Processing Unit (CPU) 169

Random Access Memory (RAM) 169

Motherboard 169

Power Supply Unit (PSU) 170

Optimizing Machine Learning Software 170

Summary 170

Chapter 10: Machine Learning Example Problems ... 171

Problem 1: Asset Price Prediction ... 171
Problem Type: Supervised Learning—Regression ... 172
Description of the Experiment ... 173
Feature Selection ... 175

Model Evaluation ... 176
Ridge Regression ... 176
Support Vector Regression (SVR) ... 178
Problem 2: Speed Dating ... 180
Problem Type: Classification ... 181
Preprocessing: Data Cleaning and Imputation ... 182

Feature Selection ... 185

Model Training and Evaluation ... 186
Method 1: Logistic Regression ... 186
Method 3: K-Nearest Neighbors (KNN) ... 189
Method 2: Bayesian Classifier ... 191

Summary ... 194

Chapter 11: Deep Learning and Other Example Problems ... 195

Autoencoders ... 195

Convolutional Neural Networks ... 202
Preprocessing ... 204

Model Building and Training ... 206
Collaborative Filtering ... 214

Summary ... 218

Chapter 12: Closing Statements ... 219

Index ... 221

About the Author

Taweh Beysolow II is a Machine Learning Scientist currently based in the United States with a passion for research and applying machine learning methods to solve problems. He has a Bachelor of Science degree in Economics from St. Johns University and a Master of Science in Applied Statistics from Fordham University. Currently, he is extremely passionate about all matters related to machine learning, data science, quantitative finance, and economics.

About the Technical Reviewer

Somil Asthana has a BTech from IITBHU India and an MS from the University of Buffalo, US, both in Computer Science. He is an Entrepreneur, Machine Learning Wizard, and BigData specialist consulting with fortune 500 companies like Sprint, Verizon, HPE, Avaya. He has a startup which provides BigData solutions and Data Strategies to Data Driven Industries in ecommerce, content / media domain.

Acknowledgments

To my family, who I am never grateful enough for. To my grandmother, from whom much was received and to whom much is owed. To my editors and other professionals who supported me through this process, no matter how small the assistance seemed. To my professors, who continue to inspire the curiosity that makes research worth pursuing. To my friends, new and old, who make life worth living and memories worth keeping. To my late friend Michael Giangrasso, who I intended on researching Deep Learning with. And finally, to my late mentor and friend Lawrence Sobol. I am forever grateful for your friendship and guidance, and continue to carry your teachings throughout my daily life.

Introduction

It is assumed that all readers have at least an elementary understanding of statistical or computer programming, specifically with respect to the R programming language. Those who do not will find it much more difficult to follow the sections of this book which give examples of code to use, and it is suggested that they return to this text upon gaining that information.

CHAPTER 1

Introduction to Deep Learning

With advances in hardware and the emergence of big data, more advanced computing methods have become increasingly popular. Increasing consumer demand for better products and companies seeking to leverage their resources more efficiently have also been leading this push. In response to these market forces, we have recently seen a renewed and widely spoken about interest in the field of machine learning. At the cross-section of statistics, mathematics, and computer science, *machine learning* refers to the science of creating and studying algorithms that improve their own behavior in an iterative manner by design. Originally, the field was devoted to developing artificial intelligence, but due to the limitations of the theory and technology that were present at the time, it became more logical to focus these algorithms on specific tasks. Most machine learning algorithms as they exist now focus on function optimization, and the solutions yielded don't always explain the underlying trends within the data nor give the inferential power that artificial intelligence was trying to get close to. As such, using machine learning algorithms often becomes a repetitive trial and error process, in which the choice of algorithm across problems yields different performance results. This is fine in some contexts, but in the case of language modeling and computer vision, it becomes problematic.

In response to some of the shortcomings of machine learning, and the significant advance in the theoretical and technological capabilities at our disposal today, deep learning has emerged and is rapidly expanding as one of the most exciting fields of science. It is being used in technologies such as self-driving cars, image recognition on social media platforms, and translation of text from one language to others. *Deep learning* is the subfield of machine learning that is devoted to building algorithms that explain and learn a high and low level of abstractions of data that traditional machine learning algorithms often cannot. The models in deep learning are often inspired by many sources of knowledge, such as game theory and neuroscience, and many of the models often mimic the basic structure of a human nervous system. As the field advances, many researchers envision a world where software isn't nearly as hard coded as it often needs to be today, allowing for a more robust, generalized solution to solving problems.

Although it originally started in a space similar to machine learning, where the primary focus was constraint satisfaction to varying degrees of complexity, deep learning has now evolved to encompass a broader definition of algorithms that are able to understand multiple levels of representation of data that correspond to different hierarchies of complexity. In other words, the algorithms not only have predictive and classification ability, but they are able to learn different levels of complexity. An example of this is found in image recognition, where a neural network builds upon recognizing eyelashes, to faces, to people, and so on. The power in this is obvious: we can reach a level of complexity necessary to create intelligent software. We see this currently in features such as autocorrect, which models the suggested corrections to patterns of speech observed, specific to each person's vocabulary.

The structure of deep learning models often is such that they have *layers* of non-linear units that process data, or neurons, and the multiple layers in these models process different levels of abstraction of the data. Figure 1-1 shows a visualization of the layers of neural networks.

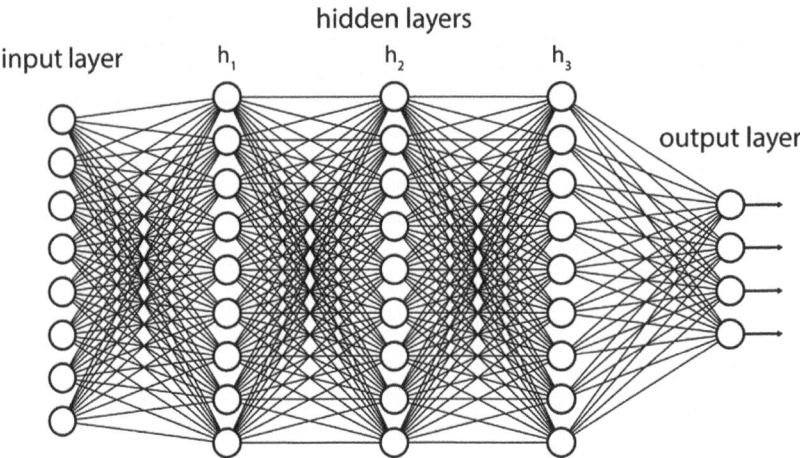

Figure 1-1. Deep neural network

Deep neural networks are distinguished by having many hidden layers, which are called "hidden" because we don't necessarily see what the inputs and outputs of these neurons are explicitly beyond knowing they are the output of the preceding layer. The addition of layers, and the functions inside the neurons of these layers, are what distinguish an individual architecture from another and establish the different use cases of a given model.

More specifically, lower levels of these models explain the "how," and the higher-levels of neural networks process the "why." The functions used in these layers are dependent on the use case, but often are customizable by the user, making them significantly more robust than the average machine learning models that are often used for classification and regression, for example. The assumption in deep learning models on a fundamental level is that the data being interpreted is generated by the interactions of different factors organized in layers. As such, having multiple layers allows the model to process the data such that it builds an understanding from simple aspects to larger constructs. The objective of these models is to perform tasks without the same degree of explicit instruction that many machine learning algorithms need. With respect to how these models are used, one of the main benefits is the promise they show when applied to unsupervised learning problems, or problems where we don't know prior to performing the experiment that the response variable y should be given a set of explanatory variables x. An example would be image recognition, particularly after a model has been trained against a given set of data. Let's say we input an image of a dog in the testing phase, implying that we don't tell the model what the picture is of. The neural network will start by recognizing eyelashes prior to a snout, prior to the shape of the dog's head, and so on until it classifies the image as that of a dog.

Deep Learning Models

Now that we have established a brief overview of deep learning, it will be useful to discuss what exactly *you* will be learning in this book, as well as describe the models we will be addressing here.

This text assumes you are relatively informed by an understanding of mathematics and statistics. Be that as it may, we will briefly review all the concepts necessary to understand linear algebra, optimization, and machine learning such that we will form a solid base of knowledge necessary for grasping deep learning. Though it does help to understand all this technical information precisely, those who don't feel comfortable with more advanced mathematics need not worry. This text is written in such a way that the reader is given all the background information necessary to research it further, if desired. However, the primary goal of this text is to show readers how to apply machine learning and deep learning models, not to give a verbose academic treatise on all the theoretical concepts discussed.

After we have sufficiently reviewed all the prerequisite mathematical and machine learning concepts, we will progress into discussing machine learning models in detail. This section describes and illustrates deep learning models.

Single Layer Perceptron Model (SLP)

The *single layer perceptron* (SLP) model is the simplest form of neural network and the basis for the more advanced models that have been developed in deep learning. Typically, we use SLP in classification problems where we need to give the data observations labels (binary or multinomial) based on inputs. The values in the input layer are directly sent to the output layer after they are multiplied by weights and a bias is added to the cumulative sum. This cumulative sum is then put into an *activation* function, which is simply a function that defines the output. When that output is above or below a user-determined threshold, the final output is determined. Researchers McCulloch-Pitts Neurons described a similar model in the 1940s (see Figure 1-2).

CHAPTER 1 ■ INTRODUCTION TO DEEP LEARNING

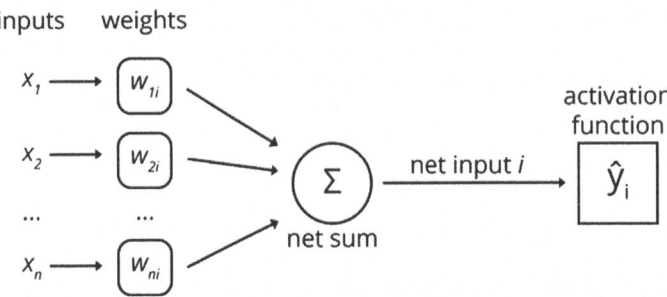

Figure 1-2. Single layer perceptron network

Multilayer Perceptron Model (MLP)

Very similar to SLP, the *multilayer perceptron* (MLP) model features multiple layers that are interconnected in such a way that they form a feed-forward neural network. Each neuron in one layer has directed connections to the neurons of a separate layer. One of the key distinguishing factors in this model and the single layer perceptron model is the back-propagation algorithm, a common method of training neural networks. Back-propagation passes the error calculated from the output layer to the input layer such that we can see each layer's contribution to the error and alter the network accordingly. Here, we use a gradient descent algorithm to determine the degree to which the weights should change upon each iteration. *Gradient descent*—another popular machine learning/optimization algorithm—is simply the derivative of a function such that we find a *scalar* (a number with magnitude as its only property) value that points in the direction of greatest momentum. By subtracting the gradient, this leads us to a solution that is more optimal than the one we currently are at until we reach a global optimum (see Figure 1-3).

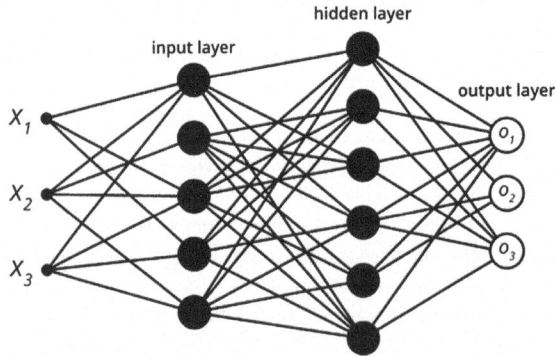

Figure 1-3. MultiLayer perceptron network

CHAPTER 1 ■ INTRODUCTION TO DEEP LEARNING

Convolutional Neural Networks (CNNs)

Convolutional neural networks (CNNs) are models that are most frequently used for image processing and computer vision. They are designed in such a way to mimic the structure of the animal visual cortex. Specifically, CNNs have neurons arranged in three dimensions: width, height, and depth. The neurons in a given layer are only connected to a small region of the prior layer. CNN models are most frequently used for image processing and computer vision (see Figure 1-4).

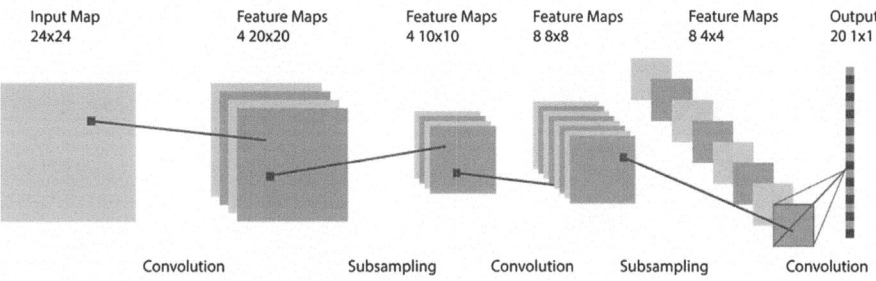

Figure 1-4. Convolutional neural network

Recurrent Neural Networks (RNNs)

Recurrent neural networks (RNNs) are models of *Artificial neural networks* (ANNs) where the connections between units form a directed cycle. Specifically, a *directed cycle* is a sequence where the walk along the vertices and edges is completely determined by the set of edges used and therefore has some semblance of a specific order. RNNs are often specifically used for speech and handwriting recognition (see Figure 1-5).

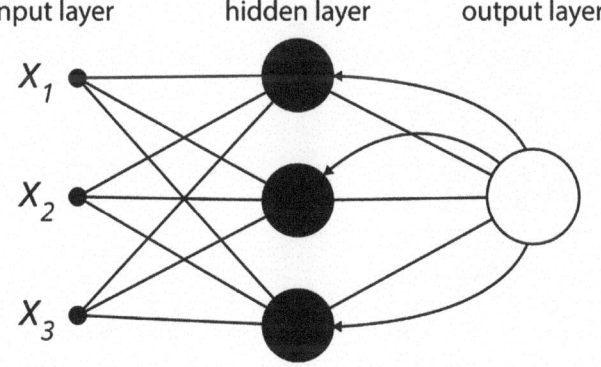

Figure 1-5. Recurrent neural network

Restricted Boltzmann Machines (RBMs)

Restricted Boltzmann machines are a type of binary Markov model that have a unique architecture, such that there are multiple layers of hidden random variables and a network of symmetrically coupled stochastic binary units. DBMs are comprised of a set of visible units and series of layers of hidden units. There are, however, no connections between units of the same layer. DMBs can learn complex and abstract internal representations in tasks such as object or speech recognition (see Figure 1-6).

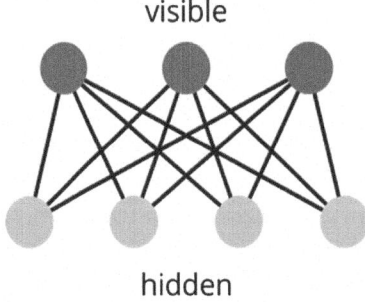

Figure 1-6. *Restricted Boltzmann machine*

Deep Belief Networks (DBNs)

Deep belief networks are similar to RBMs except each subnetwork's hidden layer is in fact the visible layer for the next subnetwork. DBNs are broadly a generative graphical model composed of multiple layers of latent variables with connections between the layers but not between the units of each individual layer (see Figure 1-7).

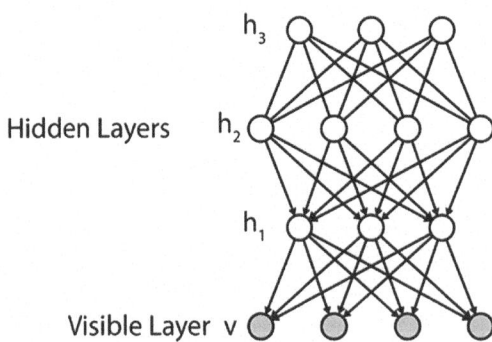

Figure 1-7. *Deep belief networks*

Other Topics Discussed

After covering all the information regarding models, we will turn to understanding the practice of data science. To aid in this effort, this section covers additional topics of interest.

Experimental Design

The emphasis of this text ultimately is to give the reader a theoretical understanding of the deep learning models such that they feel comfortable enough to apply them. As such, it is important to discuss elements of experimental design to help the reader understand proper ways to structure their research so it leads to actionable insights and not a waste of time and/or energy. Largely, I will draw upon Fisher's principles in addition to defining best practices given the problems often utilized by deep learning.

Feature Selection

A component of experimental design, but ultimately entirely a subtopic of research unto itself, I will cover the concept of variable selection and multiple methods used often by data scientists to handle high dimensional data sets. Specifically, I will speak in depth about principal components analysis as well as genetic algorithms. All the algorithms discussed are available in the R statistical language in open source packages. For those who want to research this area of research further, I'll reference papers relevant to this topic. From a deep learning perspective, we will discuss in depth how each model performs its own specific methods of feature selection by design of the layer architecture in addition to addressing recent discoveries in the field.

Applied Machine Learning and Deep Learning

For the final section of the text, I will walk the reader through using packages in the R language for machine learning and deep learning models to solve problems often seen in professional and academic settings. It is hoped that from these examples, readers will be motivated to apply machine learning and deep learning in their professional and/or academic pursuits. All the code for the examples, experiments, and research uses the R programming language and will be made available to all readers via GitHub (see the appendix for more). Among the topics discussed are regression, classification, and image recognition using deep learning models.

History of Deep Learning

Now that we have covered the general outline of the text, in addition to what the reader is expected to learn during this period, we will see how the field has evolved to this stage and get an understanding of where it seeks to go today. Although deep learning is a relatively new field, it has a rich and vibrant history filled with discovery that is still ongoing today. As for where this field finds its clearest beginnings, the discussion brings us to the 1960s.

The first working learning algorithm that is often associated with deep learning models was developed by Ivakhenenko and Lapa. They published their findings in a paper entitled "Networks Trained by the Group Method of Data Handling (GMDH)" in 1965. These were among the first deep learning systems of the feed-forward multilayer perceptron type. *Feed-forward* networks describe models where the connections between the units don't form a cycle, as they would be in a recurrent neural network. This model featured polynomial activation functions, and the layers were incrementally grown and trained by regression analysis. They were subsequently pruned with the help of a separate validation set, where regularization was used to weed out superfluous units.

In the 1980s, the neocognitron was introduced by Kunihio Fukushima. It is a multilayered artificial neural network and has primarily been used for handwritten character recognition and similar tasks that require pattern recognition. Its pattern recognition abilities gave inspiration to the convolutional neural network. Regardless, the neocognitron was inspired by a model proposed by the neurophysiologists Hubel and Wiesel. Also during this decade, Yann LeCun et al. applied the back-propagation algorithm to a deep neural network. The original purpose of this was for AT&T to recognize handwritten zip codes on mail. The advantages of this technology were significant, particularly right before the Internet and its commercialization were to occur in the late 1990s and early 2000s.

In the 1990s, the field of deep learning saw the development of a recurrent neural network that required more than 1,000 layers in an RNN unfolded in time, and the discovery that it is possible to train a network containing six fully connected layers and several hundred hidden units using what is called a wake-sleep algorithm. A *heuristic*, or an algorithm that we apply over another single or group of algorithms, a wake-sleep algorithm is a unsupervised method that allows the algorithm to adjust parameters in such a way that an optimal density estimator is outputted. The "wake" phase describes the process of the neurons firing from input to output. The connections from the inputs and outputs are modified to increase the likelihood that they replicate the correct activity in the layer below the current one. The "sleep" phase is the reverse of the wake phase, such that neurons are fired by the connections while the recognitions are modified.

As rapidly as the advancements in this field came during the early 2000s and the 2010s, the current period moving forward is being described as the watershed moment for deep learning. It is now that we are seeing the application of deep learning to a multitude of industries and fields as well as the very devoted improvement of the hardware used for these models. In the future, it is expected that the advances covered in deep learning will help to allow technology to make actions in contexts where humans often do today and where traditional machine learning algorithms have performed miserably. Although there is certainly still progress to be made, the investment made by many firms and universities to accelerate the progress is noticeable and making a significant impact on the world.

Summary

It is important for the reader to ultimately understand that no matter how sophisticated any model is that we describe here, and whatever interesting and powerful uses it may provide, there is no substitute for adequate domain knowledge in the field in which these models are being used. It is easy to fall into the trap, for both advanced and introductory practitioners, of having full faith in the outputs of the deep learning models without heavily evaluating the context in which they are used. Although seemingly self-evident, it is important to underscore the importance of carefully examining results and, more importantly, making actionable inferences where the risk of being incorrect is most limited. I hope to impress upon the reader not only the knowledge of where they can apply these models, but the reasonable limitations of the technology and research as it exists today.

This is particularly important in machine learning and deep learning because although many of these models are powerful and reach proper solutions that would be nearly impossible to do by hand, we have not determined *why* this is the case always. For example, we understand how the back-propagation algorithm works, but we can't see it operating and we don't have an understanding of what exactly happened to reach such a conclusion. The main problem that arises from this situation is that when a process breaks, we don't necessarily always have an idea as to why. Although there have been methods created to try and track the neurons and the order in which they are activated, the decision-making process for a neural network isn't always consistent, particularly across differing problems. It is my hope that the reader keeps this in mind when moving forward and evaluates this concern appropriately when necessary.

CHAPTER 2

Mathematical Review

Prior to discussing machine learning, a brief overview of statistics is necessary. Broadly, *statistics* is the analysis and collection of quantitative data with the ultimate goal of making actionable insights on this data. With that being said, although machine learning and statistics aren't the same field, they are closely related. This chapter gives a brief overview of terms relevant to our discussions later in the book.

Statistical Concepts

No discussion about statistics or machine learning would be appropriate without initially discussing the concept of probability.

Probability

Probability is the measure of the likelihood of an event. Although many machine learning models tend to be deterministic (based off of algorithmic rules) rather than probabilistic, the concept of probability is referenced specifically in algorithms such as the expectation maximization algorithm in addition to more complex deep learning architectures such a recurrent neural networks and convolutional neural networks. Mathematically, this algorithm is defined as the following:

$$\text{Probability of Event } A = \frac{\text{number of times event } A \text{ occurs}}{\text{all possible events}}$$

This method of calculating probability represents the *frequentist* view of probability, in which probability is by and large derived from the following formula. However, the other school of probability, Bayesian, takes a differing approach. Bayesian probability theory is based on the assumption that probability is conditional. In other words, the likelihood of an event is influenced by the conditions that currently exist or events that have happened prior. We define conditional probability in the following equation. The probability of an event A, given that an event B has occurred, is equal to the following:

$$P(A|B) = \frac{P(A \cap B)}{P(B)},$$

CHAPTER 2 ■ MATHEMATICAL REVIEW

$$Provided\ P(B) > 0.$$

In this equation, we read $P(A|B)$ as "the probability of A given B" and $P(A \cap B)$ as "the probability of A and B."

With this being said, calculating probability is not as simple as it might seem, in that dependency versus independency must often be evaluated. As a simple example, let's say we are evaluating the probability of two events, A and B. Let's also assume that the probability of event B occurring is dependent on A occurring. Therefore, the probability of B occurring should A not occur is 0. Mathematically, we define dependency versus independency of two events A and B as the following:

$$P(A|B) = P(A)$$

$$P(B|A) = P(B)$$

$$P(A \cap B) = P(A)P(B)$$

In Figure 2-1, we can envision events A and B as two sets, with the union of A and B as the intersection of the circles:

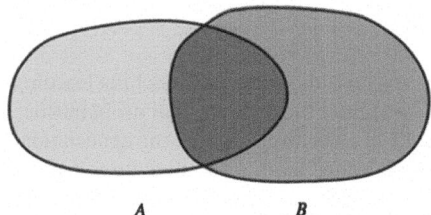

Figure 2-1. Representation of two events (A,B)

Should this equation not hold in a given circumstance, the events A and B are said to be dependent.

And vs. Or

Typically when speaking about probability—for instance, when evaluating two events A and B—probability is often in discussed in the context of "the probability of A *and* B" or "the probability of A *or* B." Intuitively, we define these probabilities as being two different events and therefore their mathematical derivations are difference. Simply stated, *or* denotes the addition of probabilities events, whereas *and* implies the multiplication of probabilities of event. The following are the equations needed:

12

And (multiplicative law of probability) is the probability of the intersection of two events A and B:

$$P(A \cap B) = P(A)P(B|A)$$

$$= P(B)P(A|B)$$

If the events are independent, then

$$P(A \cap B) = P(A)P(B)$$

Or (additive law of probability) is the probability of the union of two events A and B:

$$P(A \cup B) = P(A) + P(B) - P(A \cap B)$$

The symbol $P(A \cup B)$ means "the probability of A or B."

Figure 2-2 illustrates this.

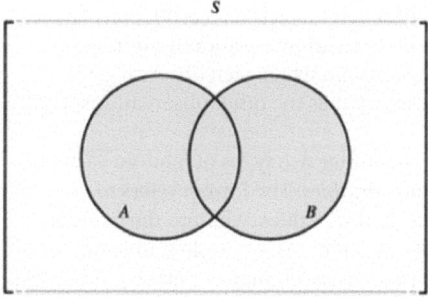

Figure 2-2. *Representation of events A,B and set S*

The probabilities of A *and* B exclusively are the section of their respective spheres which do not intersect, whereas the probability of A *or* B would be the addition of these two sections plus the intersection. We define S as the sum of all sets that we would consider in a given problem plus the space outside of these sets. The probability of S is therefore always 1.

With this being said, the space outside of A and B represents the opposite of these events. For example, say that A and B represent the probabilities of a mother coming home at 5 p.m. and a father coming home at 5 p.m. respectively. The white space represents the probability that neither of them comes home at 5 p.m.

Bayes' Theorem

As mentioned, Bayesian statistics is continually gaining appreciation within the fields of machine learning and deep learning. Although these techniques can often require considerable amounts of hard coding, their power comes from the relatively simple theoretical underpinning while being powerful and applicable in a variety of contexts. Built upon the concept of conditional probability, Bayes' theorem is the concept that the probability of an event A is related to the probability of other similar events:

$$P(B_j|A) = \frac{P(A|B_j)P(B_j)}{\sum_i^k P(A|B_i)P(B_i)}$$

Referenced in later chapters, Bayesian classifiers are built upon this formula as well as the expectation maximization algorithm.

Random Variables

Typically, when analyzing the probabilities of events, we do so within a set of random variables. We define a random variable as a quantity whose value depends on a set of possible random events, each with an associated probability. Its value is known prior to it being drawn, but it also can be defined as a function that maps from a probability space. Typically, we draw these random variables via a method know as random sampling. *Random sampling* from a population is said to be random when each observation is chosen in such a way that it is just as likely to be selected as the other observations within the population.

Broadly speaking, the reader can expect to encounter two types of random variables: *discrete random variables* and *continuous random variables*. The former refers to variables that can only assume a finite number of distinct values, whereas the latter are variables that have an infinite number of possible variables. An example is the number of cars in a garage versus the theoretical change in percentage change of a stock price. When analyzing these random variables, we typically rely on a variety of statistics that readers can expect to see frequently. But these statistics often are used directly in the algorithms either during the various steps or in the process of evaluating a given machine learning or deep learning model.

As an example, arithmetic means are directly used in algorithms such as K-means clustering while also being a theoretical underpinning of the model evaluation statistics such as mean squared error (referenced later in this chapter). Intuitively, we define the arithmetic mean as the central tendency of a discrete set of numbers—specifically it is the sum of the values divided by the number of the values. Mathematically, this equation is given by the following:

$$\bar{x} = \frac{1}{N}\sum_{i=1}^{N} x_i$$

The *arithmetic* mean, broadly speaking, represents the most likely value from a set of values within a random variable. However, this isn't the only type of mean we can use to understand a random variable. The *geometric* mean is also a statistic that describes the central tendency of a sequence of numbers, but it is acquired by using the product of the values rather than the sum. This is typically used when comparing different items within a sequence, particularly if they have multiple properties individually. The equation for the geometric mean is given as follows:

$$\left(\prod_{i=1}^{n} x_i\right)^{\frac{1}{n}} = (x_1 * x_2 * \ldots * x_n)^{\frac{1}{n}}$$

For those involved in fields where the use of time series is frequent, geometric means are useful to acquiring a measure of change over certain intervals (hours, months, years, and so on). That said, the central tendency of a random variable is not the only useful statistic for describing data. Often, we would like to analyze the degree to which the data is dispersed around the most probable value. Logically, this leads us to the discussion of variance and standard deviation. Both of these statistics are highly related, but they have a few key distinctions: *variance* is the squared value of standard deviation, and the standard deviation is often more referenced than variance across various fields. When addressing the latter distinction, this is because variance is much harder to visually describe, in addition to the fact that the units that variance is in are ambiguous. Standard deviation is in the units of the random variable being analyzed and is easy to visualize.

For example, when evaluating the efficiency of a given machine learning algorithm, we could draw the mean squared error from several epochs. It might be helpful to collect sample statistics of these variables, such that we can understand the dispersion of this statistic. Mathematically, we define variance and standard deviation as the following

Variance

$$\sigma^2 = \frac{\Sigma(X-\mu)^2}{N}$$

$$Var(X) = E\left[\left(X - E([X])\right)^2\right]$$

$$= E[X^2] - 2XE[X] + (E[X])^2$$

$$= E[X^2] - 2E[X]E[X] + (E[X])^2$$

$$= E[X^2] - 2E[X]E[X] + (E[X])^2$$

CHAPTER 2 ■ MATHEMATICAL REVIEW

Standard Deviation

$$\sigma = \sqrt{\left(\frac{\sum_{i}^{n}(x_i - \bar{x})^2}{n-1}\right)}$$

Also, covariance is useful for measuring the degree to which a change in one feature affects the other. Mathematically, we define covariance as the following:

$$cov(X,Y) = \frac{1}{n}\sum_{i=1}^{n}(x_i - \bar{x})(y_i - \bar{y})$$

Although deep learning has made significant progress in modeling relationships between variables with non-linear correlations, some estimators that one would use for more simple tasks require this as a preliminary assumption. For example, linear regression requires this to be an assumption, and although many machine learning algorithms can model complex data, some are better at it than others. As such, it is recommended that prior to selecting estimators features be examined for their relationship to one another using these prior statistics. As such, this leads us to the discussion of the correlation coefficient which measures the degree to which to variables are linearly related to each other. Mathematically, we define this as follows:

$$correlation = \rho = \frac{1}{n}\sum_{i=1}^{n}\frac{(x_i - \bar{x})(y_i - \bar{y})}{\sqrt{(x_i - \bar{x})^2(y_i - \bar{y})^2}}$$

Correlation coefficients can have a value as low as -1 and as high as 1, with the lower bound representing an *opposite* correlation and the upper bound representing *complete* correlation. A correlation coefficient of 0 represents complete lack of correlation, statistically speaking. When evaluating machine learning models, specifically those that perform regression, we typically reference the coefficient of determination (R squared) and mean squared error (MSE). We think of *R squared* as a measure of how well the estimated regression line of the model fits the distribution of the data. As such, we can state that this statistic is best known as the *degree of fitness* of a given model. MSE measures the average of the squared error of the deviations from the models predictions to the observed data. We define both respectively as the following:

Coefficient of Determination (R Squared)

$$R^2 = 1 - \sum_{i}^{n} \frac{(\hat{y}_i - y)^2}{(\hat{y}_i - \bar{y})^2}$$

Mean Squared Error (MSE)

$$MSE = \frac{1}{n}\sum_{i=1}^{n}(y_i - \bar{y})^2$$

With respect to what these values should be, I discuss that in detail later in the text. Briefly stated, though, we typically seek to have models that have high R squared values and lower MSE values than other estimators chosen.

Linear Algebra

Concepts of linear algebra are utilized heavily in machine learning, data science, and computer science. Though this is not intended to be an exhaustive review, it is appropriate for all readers to be familiar with the following concepts at a minimum.

Scalars and Vectors

A *scalar* is a value that only has one attribute: *magnitude*. A collection of scalars, known as a vector, can have both magnitude and direction. If we have more than one scalar in a given vector, we call this an *element of vector space*. Vector space is distinguished by the fact that it is sequence of scalars that can be added and multiplied, and that can have other numerical operations performed on them. *Vectors* are defined as a column vector of n numbers. When we refer to the indexing of a vector, we will describe *i* as the index value. For example, if we have a vector x, then x_1 refers to the first value in vector x. Intuitively, imagine a vector as an object similar to a file within a file cabinet. The values within this vector are the individual sheets of paper, and the vector itself is the folder that holds all these values.

Vectors are one of the primary building blocks of many of the concepts discussed in this text (see Figure 2-3). For example, in deep learning models such as Doc2Vec and Word2Vec, we typically represent words, and documents of text, as vectors. This representation allows us to condense massive amount of data into a format easy to input to neural networks to perform calculations on. From this massive reduction of dimensionality, we can determine the degree of similarity, or dissimilarity, from one document to another, or we can gain better understanding of synonyms than from simple Bayesian inference. For data that is already numeric, vectors provide an easy method of "storing" this data to be inputted into algorithms for the same purpose. The properties of vectors (and matrices), particularly with respect to mathematical operations, allow for relatively quick calculations to be performed over massive amounts of data, also presenting a computational advantage of manually operating on each individual value within a data set.

CHAPTER 2 ■ MATHEMATICAL REVIEW

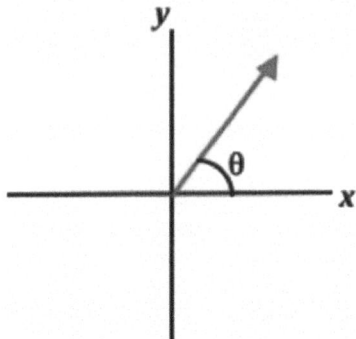

Figure 2-3. Representation of a vector

Properties of Vectors

Vector dimensions are often denoted by $\mathbb{R}^n$ or $\mathbb{R}^m$ where n and m is the number of values within a given vector. For example, $x \in \mathbb{R}^5$ denotes set of 5 vectors with real components. Although I have only discussed a column vector so far, we can also have a row vector. A transformation to change a column vector into a row vector can also be performed, known as a transposition. A *transposition* is a transformation of a matrix/vector X such that the rows of X are written as the columns of X^T and the columns of X are written as the rows of X^T.

Addition

Let's define two vectors $d = [d_1, d_2, \ldots, d_n]^T$ and $e = [e_1, e_2, \ldots, e_n]^T$ where

$$d_n = e_n, \text{ for } i = 1, 2, \ldots, n$$

The sum of the vectors is therefore the following:

$$d + e = [(d_1 + e_1), (e_2 + d_2), \ldots, (d_n + e_n)]^T$$

Subtraction

Given that the assumptions from the previous example have not changed, the difference between vectors d and e would be the following:

$$d - e = [(d_1 - e_1), (e_2 - d_2), \ldots, (d_n - e_n)]^T$$

Element Wise Multiplication

Given that the assumptions from the previous example have not changed, the product of vectors d and e would be the following:

$$d*e = \left[(d_1*e_1),(e_2*d_2),\ldots,(d_n*e_n)\right]^T$$

Axioms

Let a, b, and x be a set of vectors within set A, and e and d be scalars in B. The following axioms must hold if something is to be a vector space:

Associative Property

The associative property refers to the fact that rearranging the parentheses in a given expression will not change the final value:

$$x+(a+b)=(x+a)+b$$

Commutative Property

The commutative property refers to the fact that changing the order of the operands in a given expression will not change the final value:

$$a+b=b+a$$

Identity Element of Addition

$$a+0=a, \text{ for all } a \in A$$

Where $0 \in A$. 0 in this instance is the zero vector, or a vectors of zeros.

Inverse Elements of Addition

In this instance, for every a := A, there exists an element -a := A, which we label as the additive inverse of a:

$$a+(-a)=0$$

Identity Element of Scalar Multiplication

$$(1)a = a$$

Distributivity of Scalar Multiplication with Respect to Vector Addition

$$e(a+b) = ea + eb$$

Distributivity of Scalar Multiplication with Respect to Field Addition

$$(a+b)d = ad + bd$$

Subspaces

A *subspace* of a vector space is a nonempty subset that satisfies the requirements for a vector space, specifically that linear combinations stay in the subspace. This subset is "closed" under addition and scalar multiplication. Most notably, the zero vector will belong to every subspace. For example, the space that lies between the hyperplanes of produced by a support vector regression, a machine learning algorithm I address later, is an example of a subspace. In this subspace are acceptable values for the response variable.

Matrices

A matrix is another fundamental concept of linear algebra in our mathematical review. Simply put, a *matrix* is a rectangular array of numbers, symbols, or expressions arranged in rows and columns. Matrices have a variety of uses, but specifically are often used to store numerical data. For example, when performing image recognition with a convolutional neural network, we represent the pixels in the photos as numbers within a 3-dimensional matrix, representing the matrix for the red, green, and blue photos comprised of a color photo. Typically, we take an individual pixel to have 256 individual values, and from this mathematical interpretation an otherwise difficult-to-understand representation of data becomes possible. In relation to vectors and scalars, a matrix contains scalars for each individual value and is made up of row and column vectors. When we are indexing a given matrix A, we will be using the notation A_{ij}. We also say that $A = a_{ij}$, $A \in \mathbb{R}^{m \times n}$.

Matrix Properties

Matrices themselves share many of the same elementary properties that vectors have by definition of matrices being combinations of vectors. However, there are some key differences that are important, particularly with respect to matrix multiplication. For example, matrix multiplication is a key element of understanding how ordinary least squares regression works, and fundamentally why we would be interested in using gradient descent when performing linear regression. With that being said, the properties of matrices are discussed in the rest of this section.

Addition

Let's assume A and B are both matrices with $m \times n$ dimensions:

$$A + B = \left(A_{ij} + B_{ij}\right), \text{ for } i = 1, 2, \ldots, n$$

Scalar Multiplication

Let us assume A and B are both matrices with $m \times n$ dimensions

$$AB = \left(A_{ij} * B_{ij}\right), \text{ for } i = 1, 2, \ldots, n$$

Transposition

$$A_{ij}^T = A_{ji}$$

Types of Matrices

Matrices come in multiple forms, usually denoted by the shape that they take on. Although a matrix can take on a multitude of dimensions, there are many that will commonly references. Among the simplest is the square matrix, which is distinguished by the fact that it has an equal amount of rows and columns:

$$\mathbf{A} = \begin{bmatrix} a_{1,1} & a_{1,2} & a_{1,3} & \cdots & a_{1,n} \\ \vdots & & \ddots & & \vdots \\ a_{n,1} & a_{n,2} & a_{n,3} & \cdots & a_{n,n} \end{bmatrix}$$

It is generally unlikely that the reader will come across a square matrix, but the implications of matrix properties make discussing it necessary. That said, this brings us to discussing different types of matrices such as the diagonal and identity matrix. The *diagonal matrix* is a matrix where all the entries that are not along the main diagonal of the matrix (from the top left corner through the bottom right corner) are zero, given by the following:

$$A = \begin{matrix} 5 & 0 & 0 \\ 0 & 4 & 0 \\ 0 & 0 & 3 \end{matrix}$$

Similar to the diagonal matrix, the *identity matrix* also has zeros for values along all entries except for the diagonal of the matrix. The key distinction here, however, is that all the entries in the diagonal matrix are 1. This matrix is given by the following diagram:

$$I_n = \begin{matrix} 1 & 0 & 0 \\ 0 & 1 & 0 \\ 0 & 0 & 1 \end{matrix}$$

Another matrix you're not likely to see, but which is important from a theoretical perspective, is the symmetric matrix, whose transpose is equal to the non-transformed matrix. I describe *transpose* subsequently in this chapter, but it can be understood simply as transforming the rows into the columns and vice versa.

The final types of matrix I will define, specifically referenced in Newton's method (an optimization method described in Chapter 3), are definite and semi-definite matrices. A symmetric matrix is called positive-definite if all entries are greater than zero. But if all the values are all non-negative, the matrix is called *positive semi-definite*. Although described in greater detail in the following chapter, this is important for the purpose of understanding whether a problem has a global optimum (and therefore whether Newton's method can be used to find this global optimum).

Matrix Multiplication

Unlike vectors, matrix multiplication contains unique rules that will be helpful for readers who plan on applying this knowledge, particularly those using programming languages. For example, imagine that we have two matrices, A and B, and that we want to multiply them. These matrices can only be multiplied under the condition that the number of columns in A is the same as the number of rows in column B. We call this matrix product the *dot product* of matrices A and B. The next sections discuss examples of matrix multiplication and its products.

Scalar Multiplication

Assume we have some matrix, A, that we would like to multiply by the scalar value sigma. The result of this operation is displayed by the following diagram:

$$\sigma A = \sigma \begin{bmatrix} A_{1,1} & A_{1,2} & \cdots & A_{1,m} \\ \vdots & \ddots & & \vdots \\ A_{n,1} & A_{n,2} & \cdots & A_{n,m} \end{bmatrix} = \begin{bmatrix} \sigma A_{1,1} & \sigma A_{1,2} & \cdots & \sigma A_{1,m} \\ \vdots & \ddots & & \vdots \\ \sigma A_{n,1} & \sigma A_{n,2} & \cdots & \sigma A_{n,m} \end{bmatrix}$$

Each value in the matrix is multiplied by the scalar such in the new matrix that is subsequently yielded. Specifically, we can see this relationship displayed in the equations following related to eigendecomposition.

Matrix by Matrix Multiplication

Matrix multiplication is utilized in several regression methods, specifically OLS, ridge regression, and LASSO. It is an efficient yet simple way of representing mathematical operations on separate data sets. In the following example, let D be an $n \times m$ matrix and E be an $m \times p$ matrix such that when we multiply them both by each other, we get the following:

$$D = \begin{bmatrix} D_{1,1} & D_{1,2} & \cdots & D_{1,m} \\ \vdots & \ddots & & \vdots \\ D_{n,1} & D_{n,2} & \cdots & D_{n,m} \end{bmatrix}, E = \begin{bmatrix} E_{1,1} & E_{1,2} & \cdots & E_{1,p} \\ \vdots & \ddots & & \vdots \\ E_{m,1} & E_{m,2} & \cdots & E_{m,p} \end{bmatrix}$$

$$DE = \begin{bmatrix} DE_{1,1} & DE_{1,2} & \cdots & DE_{1,p} \\ \vdots & \ddots & & \vdots \\ DE_{n,1} & DE_{n,2} & \cdots & DE_{n,p} \end{bmatrix}$$

Assuming that the dimensions are equal, each element in one matrix is multiplied by the corresponding element of the other element, yielding a new matrix. Although walking through these examples may seem pointless, it is actually more important than it appears—particularly because all the operations will be performed by a computer. Readers should be familiar with, if only for the purpose of debugging errors in code, the products of matrix multiplication. We will see different matrix operations that also will occur in different contexts later.

Row and Column Vector Multiplication

For those wondering how exactly matrix multiplication yields a single scalar value, the following section elaborates on this further. If

$$X = \begin{pmatrix} x & y & z \end{pmatrix}, \quad Y = \begin{matrix} d \\ e \\ f \end{matrix}$$

then their matrix products are given by the following:

$$XY = \begin{pmatrix} x & y & z \end{pmatrix} \begin{matrix} d \\ e \\ f \end{matrix}$$

$$XY = xd + ye + zf$$

Contrastingly:

$$YX = \begin{matrix} d \\ e \\ f \end{matrix} \begin{pmatrix} x & y & z \end{pmatrix}$$

$$YX = \begin{matrix} dx & dy & dz \\ ex & ey & ez \\ fx & fy & fz \end{matrix}$$

Column Vector and Square Matrix

In some cases, we need to multiple a column vector by an entire matrix. In this instance, the following holds:

$$B = \begin{matrix} 1 & 2 & 3 \\ 4 & 5 & 6 \\ 7 & 8 & 9 \end{matrix}, \quad C = \begin{matrix} d \\ e \\ f \end{matrix}$$

The matrix product of B and C is given by the following:

$$YX = \begin{matrix} 1d & 2d & 3d \\ 4e & 5e & 6e \\ 7f & 8f & 9f \end{matrix}$$

Square Matrices

Among the simplest of matrix operations is when we are dealing with two square matrices, as follows:

$$B = \begin{matrix} 1 & 2 & 3 \\ 4 & 5 & 6 \\ 7 & 8 & 9 \end{matrix}, \quad D = \begin{matrix} 9 & 8 & 7 \\ 6 & 5 & 4 \\ 3 & 2 & 1 \end{matrix}$$

$$BD = \begin{matrix} 1 & 2 & 3 \\ 4 & 5 & 6 \\ 7 & 8 & 9 \end{matrix} \; x \; \begin{matrix} 9 & 8 & 7 \\ 6 & 5 & 4 \\ 3 & 2 & 1 \end{matrix}$$

$$= \begin{matrix} (1*9)+(2*6)+(3*3) & (1*8)+(2*5)+(3*2) & (1*7)+(2*4)+(3*1) \\ (4*9)+(5*6)+(6*3) & (4*8)+(5*5)+(6*2) & (4*7)+(5*4)+(6*1) \\ (7*9)+(8*6)+(9*3) & (7*8)+(8*5)+(9*2) & (7*7)+(8*4)+(9*1) \end{matrix}$$

$$BD = \begin{matrix} 30 & 24 & 18 \\ 84 & 69 & 54 \\ 138 & 114 & 90 \end{matrix}$$

By this same logic:

$$DB = \begin{matrix} 90 & 114 & 138 \\ 54 & 69 & 84 \\ 18 & 24 & 30 \end{matrix}$$

Row Vector, Square Matrix, and Column Vector

In other cases, we will perform operations on matrices/vectors with distinct shapes among each:

$$A = \begin{matrix} 9 & 8 & 7 \\ 6 & 5 & 4 \\ 3 & 2 & 1 \end{matrix}, \quad B = \begin{matrix} 1 & 2 & 3 \end{matrix}, \quad C = \begin{matrix} 4 \\ 5 \\ 6 \end{matrix}$$

$$ABC = \begin{matrix} 9 & 8 & 7 \\ 6 & 5 & 4 \\ 3 & 2 & 1 \end{matrix} \; x \; \begin{matrix} 1 & 2 & 3 \end{matrix} \; x \; \begin{matrix} 4 \\ 5 \\ 6 \end{matrix}$$

CHAPTER 2 ■ MATHEMATICAL REVIEW

$$\begin{pmatrix} 9 & 8 & 7 \\ 6 & 5 & 4 \\ 3 & 2 & 1 \end{pmatrix} X \begin{pmatrix} 4 & 10 & 18 \end{pmatrix}$$

$$ABC = \begin{pmatrix} 36 & 32 & 28 \\ 60 & 50 & 40 \\ 54 & 36 & 18 \end{pmatrix}$$

Rectangular Matrices

Our last examples address the rectangular matrix. For this example, we have two matrices Z and Y such that:

$$Z = \begin{pmatrix} 1 & 2 & 3 \\ 4 & 5 & 6 \end{pmatrix}, \quad Y = \begin{pmatrix} 9 & 8 \\ 7 & 6 \\ 5 & 4 \end{pmatrix}$$

$$ZY = \begin{pmatrix} 1 & 2 & 3 \\ 4 & 5 & 6 \end{pmatrix} x \begin{pmatrix} 9 & 8 \\ 7 & 6 \\ 5 & 4 \end{pmatrix}$$

$$= \begin{pmatrix} 9 & 40 \\ 28 & 18 \\ 10 & 240 \end{pmatrix}$$

Matrix Multiplication Properties (Two Matrices)
Not Commutative

In general, given two matrices A and B, AB ≠ BA, AB and BA may not be simultaneously defined, and even if they are, they still may not be equal. This is contrary to ordinary multiplication of numbers. For example, to specify the ordering of matrix multiplication verbally, pre-multiply A by B means BA while post-multiply A by C means AC. As long as the entries of the matrix come from a ring that has an identity and $n > 1$, there is a pair of $n \times n$ non-commuting matrices over the ring. A notable exception is the identity matrix, because it commutes with every square matrix.

CHAPTER 2 ■ MATHEMATICAL REVIEW

Distributive over Matrix Addition

Distributivity in matrices follows the same logic as it does in vectors. As such, the following axioms hold:
Left distributivity:

$$A(B+C) = AB + BC$$

Right distributivity:

$$(A+B)C = AC + BC$$

Index notation of these operations respectively are the following:

$$\sum_k A_{ik}(B_{kj} + C_{kj}) = \sum_k A_{ik}B_{kj} + \sum_k A_{ik}C_{kj}$$

$$\sum_k (A_{ik} + B_{ik})C_{kj} = \sum_k A_{ik}C_{kj} + \sum_k B_{ik}C_{kj}$$

Scalar Multiplication Is Compatible with Matrix Multiplication

Following our discussion earlier of scalar multiplication with respect to a matrix, we see here that distributivity of scalar multiplication with matrices also holds. For example, we have the following equation, which proves this as such:

$$\lambda(AB) = (\lambda A)B$$

$$(AB)\lambda = A(B\lambda)$$

λ is a scalar. If the entries of the matrix are real or complex numbers, then all four quantities are equal. More generally, all four are equal if lambda belongs to the center of the ring of entries of the matrix, because in this case $\lambda X = X\lambda$.
Index notation of this is the following:

$$\lambda \sum_K (A_{ik}B_{kj}) = \sum_k (\lambda A_{ik})B_{kj} = \sum A_{ik}(\lambda B_{kj})$$

$$\sum_k (A_{ik}B_{kj})\lambda = \sum (A_{ik}\lambda)B_{kj} = \sum_k A_{ik}(B_{kj}\lambda)$$

27

Transpose

As referred to earlier, the *transpose* of a matrix is an operation on a matrix where the product of this transformation is a new matrix in which the new matrix's rows are the original matrix's columns and the new matrix's columns are the original matrix's rows. The following equation shows how we denote this transformation, given two matrices A and B

$$(AB)^T = B^T A^T$$

where T denotes the transpose, the interchange of row I with column I in a matrix. This identity holds for any matrices over a commutative ring, but not for all rings in general. Note that A and B are reversed.

Index notation:

$$\left[(AB)^T\right]_{ij} = (AB)_{ji}$$

$$= \Sigma_K (A)_{jk} (B)_{ki}$$

$$= \Sigma_k (A^T)_{kj} (B^T)_{ik}$$

$$= \Sigma_k (B^T)_{ik} (A^T)_{kj}$$

$$= \left[(B^T)(A^T)\right]_{ij}$$

Trace

The *trace* of a product AB is independent of the order of A and B. The trace can also be thought of as the diagonal of a matrix:

$$tr(AB) = tr(BA)$$

Index notation:

$$tr(AB) = \Sigma_i \Sigma_k A_{ik} B_{ki}$$

$$= \Sigma_k \Sigma_i B_{ki} A_{ik}$$

$$= tr(BA)$$

Norms

A *norm* is a function that assigns a strictly positive length or size to each vector in a vector space. In machine learning you will encounter many various norms, and they play a vital role in reducing the MSE of regression models in addition to increasing accuracy in classification models. For example, ridge regression uses an L2 norm to shrink the regression coefficients during periods of high multicollinearity, and LASSO uses an L1 norm to shrink some regression coefficients to zero. I will review both of these regression models in detail in Chapter 3.

In the context of deep learning, experimentation of adding different layers in deep neural networks, in which norms are used to perform dimensionality reduction on data, have proved successful at some tasks. For example, use of an L2 norm layer was performed in a convolutional neural network. But this also can be used as a dissimilarity/loss measure in multilayer perceptrons rather than a traditional gradient function.

Euclidean Norm

This describes the distance over a vector within Euclidean space in $\mathbb{R}^n$. Let's assume $x = (x_1, x_2, \ldots, x_n)$.

L2 Norm

This gives the distance from the origin point within the vector to the last point within x, and is often referred to as the L2 norm:

$$\|x\|_2^2 = \sqrt{x_1^2 + x_2^2 + \ldots + x_n^2}$$

L1 Norm

This is the same equation as the L2 norm except that the scalars are not squared:

$$\|x\| = \sqrt{x_1 + x_2 + \ldots + x_n}$$

The shape of the L1 verses L2 norms are distinguished as shown in Figure 2-4.

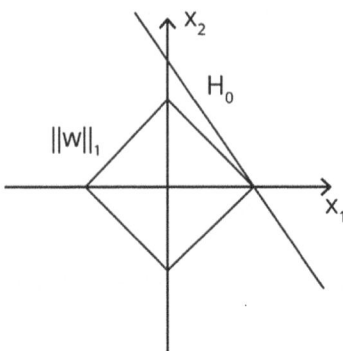 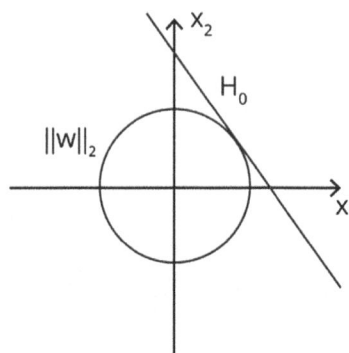

Figure 2-4. L1 and L2 norm shapes

Note that with the L1 norm we observe a square (or cubic) shape, and with the L2 norm we observe a circle (or spherical) shape. In certain situations, it's optimal to use the L1 norm to perform variable selection at the same time while also performing regression analysis, but this is an issue that isn't always present, and I discuss it in further detail in Chapter 8.

The advantage of using an L1 norm is obvious in that you can perform feature selection while performing regression. However, it should be noted that performing feature selection after reduction of the data set has already occurred can encourage overfitting. Strategies and general practices for building a robust model are reviewed more extensively in Chapter 8, but it is generally suggested that readers use the L1 norm in the instance that there has been little to no feature selection performed prior to fitting the data to the model.

For those interested in vehicle routing problems, the *taxicab (Manhattan) norm* is relevant for those who want to focus on fields related to transportation and/or delivery or packages/persons. The taxicab norm describes the distance a taxicab would travel along a given city block:

$$\|x\| = \sum_i |x_i|, \text{ for } i = 1, 2, \ldots, n$$

The absolute value norm is a norm on the one-dimensional vector spaces formed by real or complex numbers. Absolute value norms have been used in place of other loss functions or dissimilarity functions:

$$\|x\| = |x|$$

P-norm

Let $p \geq 1$ be a real number:

CHAPTER 2 ■ MATHEMATICAL REVIEW

$$\|x\|_p = \Sigma \left(|x_i|^p\right)^{\frac{1}{p}}$$

The shape of this norm is shown in Figure 2-5.

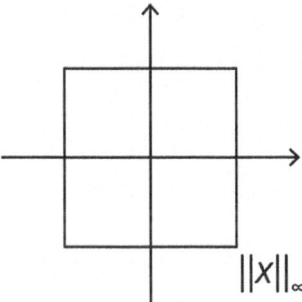

Figure 2-5. *P-norm*

For $p=1$, we get the taxicab norm, for $p=2$, we get the Euclidean norm, and as $p \to \infty$, we get the infinity norm or the maximum norm. The p-norm is related to the generalized mean or power mean. When $0 < p < 1$, though, we don't get a discretely defined norm, because it violates the triangle inequality. The *triangle inequality* states that any given side of a triangle must be less than or equal to the sum of the other two sides.

Matrix Norms

A matrix norm is a function from $\mathbb{R}^{n \times n} \to \mathbb{R}$ that satisfies a given number of properties, symbolized by $\|A\|$ given a matrix A.

The properties are as follows:

1. $\|A\| > 0$ *for all* $M \in \mathbb{R}^{n \times n}$ *and also* $\|A\| = 0$ *if* $A = 0$

2. $\|aM\| = |a| * \|M\|$ *for all* $a \in \mathbb{R}^n$

3. $\|M + N\| \le \|M\| + \|N\|$

4. $\|MN\| \le \|M\| * \|N\|$

Inner Products

An important type of vector space that's referenced often in machine learning literature is the inner product. This element of vector space allows someone to know the length of a vector or the angle between two vectors. In addition, you can also determine from the inner product normed vector space. Specifically, the *inner product* is the function utilized in the kernels of support vector machines to compute the images of the data that the support vector machine puts into feature space from the input space. The inner product space of is a function $\langle .,. \rangle$ defined by the following, where u and v are vectors, $u = [u_1, u_2, \ldots, u_n], v = [v_1, v_2, \ldots, v_n]$:

$$\langle u, v \rangle = u_1 v_1 + u_2 v_2 + \ldots + u_n v_n \text{ for } i = 1, 2, \ldots, n$$

For a function to be an inner product, it must satisfy three axioms:
Conjugate symmetry:

$$\langle u, v \rangle = \overline{\langle v, u \rangle}$$

Linearity in the first argument:

$$\langle au + bv, w \rangle = a \langle u, w \rangle + b \langle v, w \rangle$$

Positive-definiteness:

$$\text{For any } u \in V, \langle u, u \rangle \geq 0; \text{ and } \langle u, u \rangle = 0 \text{ only if } u = 0$$

Norms on Inner Product Spaces

Inner product spaces naturally have a defined norm, which is based upon the norm of the space itself, given by the following:

$$\|\langle x \rangle\| = \sqrt{\langle x \rangle}$$

Directly from the axioms, we can prove the following: The Cauchy-Schwartz inequality states that for all vectors u and v of an inner product space, the following is true:

$$\|\langle u, v \rangle\|^2 \leq \langle u, u \rangle * \langle v, v \rangle$$

$$|\langle u, v \rangle| \leq \|u\| * \|v\|$$

The two sides are only considered equal if and only if u and v are linearly dependent, which means that they would have to be parallel, one of the vectors has a magnitude of zero, or one is a scalar multiplier of the other.

Proofs

First proof: expanding out the brackets and collecting together identical terms yields the following equation:

$$\Sigma_i^n \Sigma_j^n (a_i b_j - a_j b_i)^2 = \left(\Sigma_i^n a_i^2\right)\Sigma_j^n b_j^2 + \left(\Sigma_i^n b_i^2\right)\Sigma_j^n a_j^2 - 2\left(\Sigma a_i b_i\right)\Sigma_j^n b_j a_j$$

$$= 2\left(\Sigma_i^n a_i^2\right)\left(\Sigma_i^n b_i^2\right) - 2\left(\Sigma_i^n a_i b_i\right)^2$$

Because the lefthand side of the equation is the sum of squares of real numbers, it is greater than or equal to zero. As such, the following must be true:

$$\left(\Sigma_i^n a_i^2\right)\left(\Sigma_i^n b_i^2\right) \geq \left(\Sigma_i^n a_i b_i\right)^2$$

Second proof: consider the following quadratic polynomial equation:

$$f(x) = \left(\Sigma_i^n a_i^2\right)x^2 - 2\left(\Sigma_i^n a_i b_i\right)x + \Sigma_i^n b_i^2 = \Sigma(a_i x - b_i)^2$$

Because $f(x) \geq 0$ *for* any $x \in \mathbb{R}$, it follows that the discriminant of $f(x)$ is negative, and therefore the following must be the case:

$$\left(\Sigma_i^n a_i b_i\right)^2 - \left(\Sigma_i^n a_i^2\right)\left(\Sigma_i^n b_i^2\right) \leq 0$$

Third proof: consider the following two Euclidean norms A and B:

$$\text{Let } A = \sqrt{a_1^2 + a_2^2 + \ldots + a_n^2}, \quad B = \sqrt{b_1^2 + b_2^2 + \ldots + b_n^2}$$

By the arithmetic-geometric means inequality, we have

$$\frac{\Sigma_i^n (a_i b_i)}{AB} \leq \Sigma_i^n \left(\frac{1}{2}\right)\left(\left(\frac{a_i^2}{A^2}\right) + \left(\frac{b_i^2}{B^2}\right)\right) = 1,$$

such that

$$\Sigma a_i b_i \leq AB = \sqrt{a_1^2 + a_2^2 + \ldots + a_n^2}\sqrt{b_1^2 + b_2^2 + \ldots + b_n^2}$$

Thus, the following is yielded:

$$\left(\Sigma a_i b_i\right)^2 \leq \left(\Sigma_i^n a_i^2\right)\left(\Sigma_i^n b_i^2\right)$$

Orthogonality

Orthogonality is described as a measure or degree of unrelatedness. For example, an orthogonal transformation of a vector yields a vector such that it is unrelated to the vector we transformed. The geometric interpretation of the inner product in terms of angle and length motivates much of the terminology we use in regard to those spaces. Indeed, an immediate consequence of the Cauchy-Schwarz inequality is that it justifies defining the angle between two non-zero vectors:

$$\text{Angle}(x,y) = \arccos\frac{\langle x,y \rangle}{\|x\|*\|y\|}$$

Outer Product

The tensor product of two vectors is related slightly to the inner product previously defined. A *tensor* product is a way of creating a new vector space analogous to multiplication of integers:

Let u and v equal two vectors where $x = [x_1, x_2, x_3]$, $y = [y_1, y_2, y_3]^T$

$$y \otimes x = yx^T = \begin{matrix} y_1 \\ y_2 \\ y_3 \end{matrix} * \begin{matrix} x_1 & x_2 & x_3 \end{matrix} = \begin{matrix} y_1 x_1 & y_1 x_2 & y_1 x_3 \\ y_2 x_1 & y_2 x_2 & y_2 x_3 \\ y_3 x_1 & y_3 x_2 & y_3 x_3 \end{matrix}$$

Eigenvalues and Eigenvectors

An *eigenvalue* is a number derived from a square matrix, which corresponds to a specific *eigenvector*, also associated with a square matrix. Together, they "provide the eigendecomposition of a matrix." Plainly spoken, the eigendecomposition of a matrix merely provides the matrix in the form of eigenvectors and their corresponding eigenvalues. Eigendecomposition is important because it is a "method by which we can find the maximum (or minimum) of functions involving matrices."

CHAPTER 2 ■ MATHEMATICAL REVIEW

Eigendecomposition:

$$Au = \lambda u$$

$$(A - \lambda I)u = 0$$

Where A = square matrix, and u = eigenvector to matrix A (if length of vector changes when multiplied by A):

$$\lambda = \text{eigenvalue to corresponding eigenvecvtor u}$$

Assume the following is also true:

$$A = \begin{pmatrix} 2 & 3 \\ 2 & 1 \end{pmatrix}$$

Therefore:

$$u_1 = \begin{pmatrix} 3 \\ 2 \end{pmatrix}, \; u_2 = \begin{pmatrix} -\frac{1}{1} \end{pmatrix}, \; \lambda_1 = 4, \; \lambda_2 = -1$$

For most applications, the eigenvectors are normalized to a unit vector as such:

$$u^T u = 1$$

Eigenvectors of A furthermore are put together in a matrix U. Each column of U is an eigenvector of A. The eigenvalues are stored in a diagonal matrix $\wedge$, where the trace, or diagonal, of the matrix gives the eigenvalues. Thus we rewrite the first equation accordingly:

$$AU = U\wedge$$

$$A = U \wedge U^{-1}$$

$$= \begin{bmatrix} 3 & -1 \\ 2 & 1 \end{bmatrix} \begin{bmatrix} 4 & 0 \\ 0 & -1 \end{bmatrix} \begin{bmatrix} 2 & 2 \\ -4 & 6 \end{bmatrix}$$

$$= \begin{bmatrix} 2 & 3 \\ 2 & 1 \end{bmatrix}$$

A graphical representation of eigenvectors is given in Figure 2-6.

CHAPTER 2 ■ MATHEMATICAL REVIEW

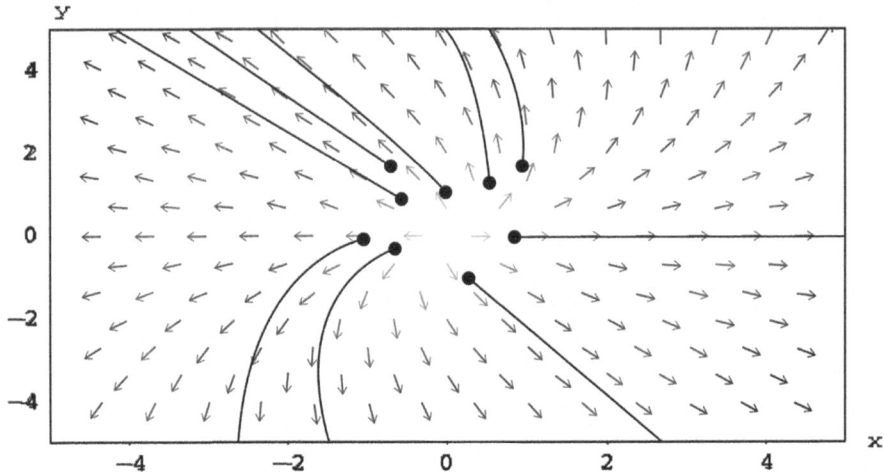

Figure 2-6. *Visulaization of eigenvectors*

Eigenvectors and eigenvalues become an integral part of understanding a technique discussed later in our discussion regarding a variable selection technique called principal components analysis (PCA). The eigendecomposition of a symmetric positive semi-definite matrix yields an orthogonal basis of eigenvectors, each of which has a non-negative eigenvalue. PCA studies linear relations among variables and is performed on the covariance matrix, or the correlation matrix, of the input data set. For the covariance or correlation matrix, the eigenvectors correspond to principal components and the eigenvalues to the variance explained by the principal components. Principal component analysis of the correlation matrix provides an orthonormal eigenbasis for the space of the observed data: in this basis, the largest eigenvalues correspond to the principal components that are associated with containing the most covariability of the observed dataset.

Linear Transformations

A linear transformation is a mapping $V \to W$ between two modules that preserves the operations of addition and scalar multiplication. When $V = W$, we call this a linear operator, or endomorphism, of V. Linear transformations always map linear subspaces onto linear subspaces, and sometimes this can be in a lower dimension. These linear maps can be represented as matrices, such as rotations and reflections. An example of where linear transformations are used is specifically PCA. Discussed in detail later, PCA is an orthogonal linear transformation of the features in a data set into uncorrelated principal components such that for K features, we have K principal components. I discuss orthogonality in detail in the following sections, but for now I focus on the broader aspects of PCA. Each principal component retains the variance from the original data set but gives us a representation of it such that we can infer the importance of a given principal component based on the contribution of the variance to the data set it provides. When translating this to the original data set, we then can remove features from the data set that we feel don't exhibit significant amounts of variance.

A function $\mathcal{L}: \mathbb{R}^n \to \mathbb{R}^m$ is called a linear transformation if the following is true:

$$\mathcal{L}(ax) = a\mathcal{L}(x) \text{ for every } x \in \mathbb{R}^n \text{ and } a \in \mathbb{R}$$

$$\mathcal{L}(x+y) = \mathcal{L}(x) + \mathcal{L}(y) \text{ for every } x, y, \in \mathbb{R}^n$$

When we fix the bases for $\mathbb{R}^n$ and $\mathbb{R}^m$, the linear transformation $\mathcal{L}$ can be represented by a matrix A. Specifically, there exists $A \in \mathbb{R}^{m \times n}$ such that the following representation holds. Suppose $x \in \mathbb{R}^n$ is a given vector and x' is the representative of x with respect to the given basis for $\mathbb{R}^m$. If $y = \mathcal{L}(x)$ and Y' is the representative of y with respect to the given basis for $\mathbb{R}^m$, then

$$y' = Ax'$$

We call A the matrix representation of $\mathcal{L}$ with respect to the given bases for $\mathbb{R}^n$ and $\mathbb{R}^m$.

Quadratic Forms

A *quadratic form* is a homogenous polynomial of the second degree in a number of variables and have applications in machine learning. Specifically, functions we seek to optimize that are twice differentiable can be optimized using Newton's method. The power in this is that if a function is twice differentiable, we know that we can reach an objective minimum.

A quadratic form $f: \mathbb{R}^n \to \mathbb{R}^m$ is a function such that the following holds true:

$$F(x) = x^T Q x$$

Where Q is an $n \times n$ real matrix. There is no loss of generality in assuming Q to be symmetric—that is, $Q = Q^T$.

Minors of a matrix Q are the determinants of the matrices obtained by successively removing rows and columns from Q. The principal minors are detQ itself and the determinants of matrices obtained by removing an ith row and an ith column.

Sylvester's Criterion

Sylvester's criterion is necessary and sufficient to determine whether a matrix is positive semi-definite. Simply, it states that for a matrix to be positive semi-definite, all the leading principal minors must be positive.

Proof: if real-symmetric matrix A has non-negative eigenvalues that are positive, it is called positive-definite. When the eigenvalues are just non-negative, A is said to be positive semi-definite.

CHAPTER 2 ■ MATHEMATICAL REVIEW

A real-symmetric matrix A has non-negative eigenvalues if and only if A can be factored as $A = B^T B$, and all eigenvalues are positive if and only if B is non-singular.

Forward implication: if $A \in R^{n \times n}$ is symmetric, then there is an orthogonal matrix P such that $A = PDP^T$, where $D = \text{diag}(\lambda_1, \lambda_2, \ldots, \lambda_n)$ is a real diagonal matrix with entries such that its columns are the eigenvectors of A. If $\lambda_i \geq 0$ for each I, $D^{\frac{1}{2}}$ exists.

Reverse implication: if A can be factored as $A = B^\wedge TB$, then all eigenvalues of A are non-negative because for any eigenpair (x, λ)

$$\lambda = \left(\frac{x^T A x}{x^{Tx}}\right) = \left(\frac{x^T B^T B x}{x^T x}\right) = \left(\frac{\|Bx\|^2}{\|x\|^2}\right) \geq 0$$

Orthogonal Projections

A *projection* is linear transformation P from a vector space to itself such that $P^2 = P$. Intuitively, this means that whenever P is applied twice to any value, it gives the same result as it it were applied once. Its image is unchanged and this definition generalizes the idea of graphical projection moreover. $\mathcal{V}$ is a subspace of R^n if $x_1, x_2 \in \mathcal{V} \to \alpha x_1 + \beta x_2 \in \mathcal{V}$ for all $\alpha, \beta \in \mathbb{R}$. The dimension of this subspace is also equal o the maximum number of linearly independent vectors in $\mathcal{V}$. If $\mathcal{V}$ is a subspace of R^n, the orthogonal complement of $\mathcal{V}$, demoted $\mathcal{V}^\perp$, consists of all vectors that are orthogonal to every vector in $\mathcal{V}$. Thus, the following is true:

$$\mathcal{V}^\perp = \{x : v^T x = 0 \text{ for all } v \in \mathcal{V}\}$$

The orthogonal complement of $\mathcal{V}$ is also a subspace. Together, $\mathcal{V}$ and $\mathcal{V}^\perp$ span R^n in the sense that every vector $x \in \mathbb{R}^n$ can be represented as

$$x = x_1 + x_2$$

where $x_1 \in \mathcal{V}$ and $x_2 \in \mathcal{V}^\perp$. We call the above representation the *orthogonal decomposition* of x with respect to $\mathcal{V}$. We say that x_1 and x_2 are orthogonal projections of x onto the subspaces $\mathcal{V}$ and $\mathcal{V}^\perp$ respectively. We write $\mathbb{R}^n = \mathcal{V} \otimes \mathcal{V}^\perp$, and say that $\mathbb{R}^n$ is a direct sum of $\mathcal{V}$ and $\mathcal{V}^\perp$. We say that a linear transformation of P is an orthogonal projector onto $\mathcal{V}$ for all $x \in \mathbb{R}^n$, we have $Px \in \mathcal{V}$ and $x - Px \in \mathcal{V}^\perp$.

Range of a Matrix

The *range* of a matrix defines the number of column vectors it contains.
Let $A \in \mathbb{R}^{m \times n}$. The range, or image, of A, is written as the following:

$$\mathcal{R}(A) \triangleq \{Ax : x \in \mathbb{R}^n\}$$

CHAPTER 2 ■ MATHEMATICAL REVIEW

Nullspace of a Matrix

The nullspace of a linear map $\mathcal{L}: \mathcal{V} \to \mathcal{W}$ between two vector spaces is the set of all elements of v of $\mathcal{V}$ for which $\mathcal{L}(v) = 0$, where zero denotes the zero vector in $\mathcal{W}$.

The nullspace, or kernel, of A is written as the following:

$$\mathcal{N}(A) \triangleq \{x \in \mathbb{R}^n : Ax = 0\}$$

Hyperplanes

Earlier I mentioned the significance of the support vector machine and the hyperplane. In the context of regression problems, the observations within the hyperplane are acceptable as response variable solutions. In the context of classification problems, the hyperplanes form the boundaries between different classes of observations (shown in Figure 2-7).

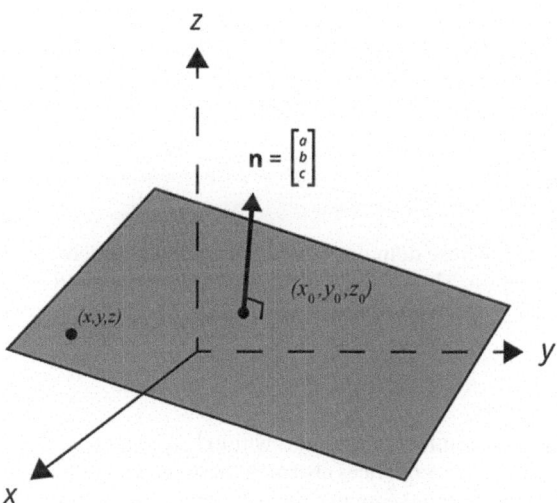

Figure 2-7. Visualization of hyperplane

We define a *hyperplane* as a subspace of one dimension less than its ambient space, otherwise known as the feature space surrounding the object.

Let $u = [u_1, u_2, \ldots, u_n]$, $u \in \mathbb{R}$, where at least one of the u_i is non-zero. The set of all points $x = [x_1, x_2, \ldots, x_n]^T$ that satisfy the linear equation

$$u_1 x_1 + u_2 x_2 + \ldots + u_n x_n = v$$

is called a hyperplane of the space $\mathbb{R}^n$. We may describe the hyperplane with the following equation:

$$\{x \in \mathbb{R}^n : u^T x = v\}$$

A hyperplane is not necessarily a subspace of $\mathbb{R}^n$ because, in general, it does not contain the origin. For $n = 2$, the equation of the hyperplane has the form $u_1 x_1 + u_2 x_2 = v$, which is the equation of a straight line. Thus, straight lines are hyperplanes in $\mathbb{R}^2$. In $\mathbb{R}^3$, hyperplanes are ordinary planes. The hyperplane H divides $\mathbb{R}^n$ into two half spaces, denoted by the following:

$$H_+ = \{x \in \mathbb{R}^n : u^T x \geq 0\},$$

$$H_- = \{x \in \mathbb{R}^n : u^T x \leq 0\}.$$

Here H_+ is the positive half-space, and H_- is the negative half-space. The hyperplane H itself consists of the points for which $\langle u, x - a \rangle = 0$, where $a = [a_1, a_2, \ldots, a_n]^T$ is an arbitrary point of the hyperplane. Simply stated, the hyperplane H is all of the points x for which the vectors u and x - a are orthogonal to one another.

Sequences

A *sequence* of real numbers is a function whose domain is the set of natural numbers 1,2,...,k, and whose range is contained in $\mathbb{R}$. Thus, a sequence of real numbers can be viewed as a set of numbers $\{x_1, x_2, \ldots, x_k\}$, which is often also denoted as $\{x_k\}$.

Properties of Sequences

The *length* of a sequence is defined as the number of elements within it. A sequence of finite length n is also called an n-tuple. *Finite* sequences also include sequences that are empty or ones that have no elements. An *infinite* sequence refers to a sequence that is infinite in one direction. It is therefore described as having a first element, but not having a final element. A sequence with neither a first nor a final element is known as a *two-way infinite* sequence or *bi-infinite* sequence.

Moreover, a sequence is said to be monotonically increasing if each term is greater than or equal to the one before it. For example, the sequence $an(n) = 1$ is monotonically increasing if an only if for all $a_{n+1} \geq a_n$. The terms *non-decreasing* and *non-increasing* are often used in place of increasing and decreasing in order to avoid any possible confusion with strictly increasing and strictly decreasing respectively.

If the sequence of real number is such that all the terms are less than some real numbers, then the sequence is said to be bounded from above. This means that there exists M such that for all n, $a_n \leq M$. Any such M is called an upper bound. Likewise, if, for some real m, $a_n \geq m$ for all n greater than some N, then the sequence is bounded from below, and any such m is called the lower bound.

Limits

A *limit* is the value that a function or sequence approaches as the input or index approaches some value. A number $x^* \in \mathbb{R}$ is called the limit of the sequence if for any positive ϵ there is a number K such that for all $k > K, |xk - x^*| < \epsilon$:

$$x^* = \lim_{k \to \infty} xk$$

A sequence that has a limit is called a *convergent* sequence. Informally speaking, a singly infinite sequence has a limit, if it approaches some value L, called the limit, as n becomes very large. If it converges towards some limit, then it is *convergent*. Otherwise it is *divergent*. Figure 2-8 shows a sequence converging upon a limit.

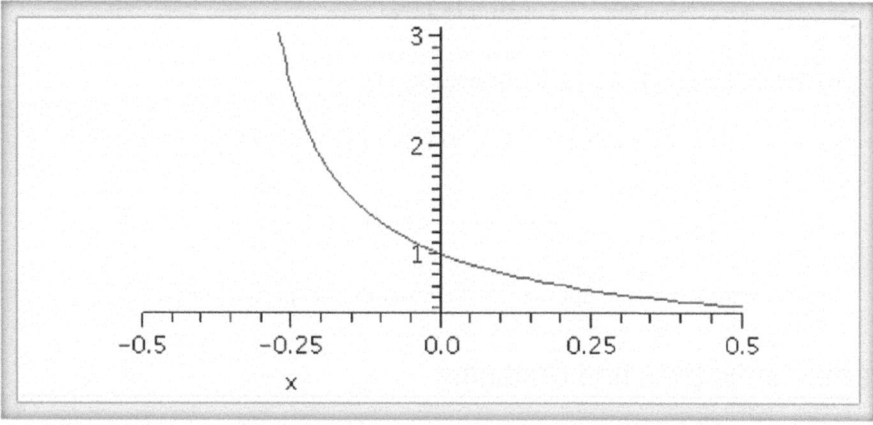

Figure 2-8. *A function converging upon 0 as x increases*

We typically speak of convergence within the context of machine learning and deep learning with reaching an *optimal solution*. This is ultimately the goal of all of our algorithms, but this becomes more ambiguous with the more difficult use cases readers encounter. Not every solution has a single global optimum—instead it could have local optima. Methods of avoiding these local optima are more specifically addressed in later chapters. Typically this requires parameter tuning of machine learning and deep learning algorithms, the most difficult part of the algorithm training process.

Derivatives and Differentiability

Differentiability becomes an important part of machine learning and deep learning, most specifically for the purpose of parameter updating. This can be seen via the backpropagation algorithm used to train multilayer perceptrons and the parameter updating of convolutional neural networks and recurrent neural networks. A derivative of a function measures the degree of change in one quantity to the degree of another. One of the most common examples of a derivative is a slope (change in y over x), or the return of a stock (price percentage change over time). This is a fundamental tool for calculus but is also the basis of many of the models we will study in the latter part of the book.

A function is considered to be *affine* if there exists a linear function $\mathcal{L}: \mathbb{R}^n \to \mathbb{R}^m$ and a vector $y \in \mathbb{R}^m$ such that

$$A(x) = \mathcal{L}(x) + y$$

for every $x \in \mathbb{R}^n$. Consider a function $f: \mathbb{R}^n \to \mathbb{R}^m$ and a point $x_0 \in \mathbb{R}^n$. We want to find an affine function A that approximates f near the point x0. First, it's natural to impose this condition:

$$A(x_0) = f(x_0)$$

We obtain $y = f(x_0) - \mathcal{L}(x_0)$. By the linearity of L,

$$\mathcal{L} + y = \mathcal{L}(x) - \mathcal{L}(x_0) + f(x_0) = \mathcal{L}(x - x_0) + f(x_0)$$

$$A(x) = \mathcal{L}(x - x_0) + f(x_0)$$

We also require that $A(x)$ approaches $f(x)$ faster than x approaches x_0.

Partial Derivatives and Gradients

Also utilized heavily in various machine learning derivations is the *partial derivative*. It is similar to a derivative, except we only take the derivative of one of the variables in the function and hold the others constant, whereas in a total derivative all the variables are evaluated. The gradient descent algorithm is discussed in Chapter 3, but we can discuss the broader concept of the gradient itself now. A *gradient* is the generalization of the concept of a derivative when applied to functions of several variables. The gradient represents the point of greatest rate of increase in the function, and its magnitude is the slope of the graph in that direction. It's a vector field whose components in a coordinate system will transform when going from one system to another:

$$\nabla f(x) = \text{grad } f(x) = \frac{df(x)}{dx}$$

Hessian Matrix

Functions can be differentiable more than once, which leads us to the concept of the Hessian matrix. The *Hessian* is a square matrix of second-order partial derivatives of a scalar values function, or scalar field:

$$\mathbf{H} = \begin{pmatrix} \frac{\partial^2 f}{\partial x_1^2} & \frac{\partial^2 f}{\partial x_1 \partial x_2} & \cdots & \frac{\partial^2 f}{\partial x_1 \partial x_n} \\ \vdots & \ddots & & \vdots \\ \frac{\partial^2 f}{\partial x_n \partial x_1} & \frac{\partial^2 f}{\partial x_n \partial x_2} & \cdots & \frac{\partial^2 f}{\partial x_n^2} \end{pmatrix}$$

If the gradient of a function is zero at some point x, then f has a critical point at x. The determinant of the Hessian at x is then called the *discriminant*. If this determinant is zero, then x is called a degenerate critical point of f, or a non-Morse critical point of f. Otherwise, it is non-degenerate.

A Jacobian matrix is the matrix of first-order partial derivatives of a vector values function. When this is a square matrix, both the matrix and its determinant are referred to as the *Jacobian*:

$$\mathbf{J} = \frac{\mathrm{d}f}{\mathrm{d}x} = \begin{bmatrix} \frac{\partial f}{\partial x_1} \cdots \frac{\partial f}{\partial x_n} \end{bmatrix} = \begin{bmatrix} \frac{\partial f1}{\partial x_1} & \cdots & \frac{\partial f_1}{\partial x_n} \\ \vdots & \ddots & \vdots \\ \frac{\partial f_m}{\partial x_1} & \cdots & \frac{\partial f_m}{\partial x_n} \end{bmatrix}$$

Summary

This brings us to the conclusion of the basic statistics and mathematical concepts that will be referenced in later chapters. Readers should feel encouraged to check back with this chapter when unsure about anything in later chapters. Moving forward, we'll address the more advanced optimization techniques that power machine learning algorithms, as well as those same machine learning algorithms that formed the inspiration of the deep learning methods we'll tackle afterwards.

CHAPTER 3

A Review of Optimization and Machine Learning

Before we dive into the models and components of deep learning in depth, it's important to address the broader field it fits into, which is machine learning. But before that, I want to discuss, if only briefly, optimization. *Optimization* refers to the selection of a best element from some set of available alternatives. The objective of most machine learning algorithms is to find the optimal solution given a function with some set of inputs. As already mentioned, this often comes within the concept of a supervised learning problem or an unsupervised learning problem, though the procedures are roughly the same.

Unconstrained Optimization

Unconstrained optimization refers to a problem in which we much reach an optimal solution. In contrast to constrain optimization, there are constraints placed on what value of x we choose, allowing us to approach the solution from significantly more avenues. An example of an unconstrained optimization problem is the following toy problem:

$$\text{Minimize}ف(x), \text{ where } f(x) = x^2, x \in [-100, 100]$$

Figure 3-1 visualizes this function.

CHAPTER 3 A REVIEW OF OPTIMIZATION AND MACHINE LEARNING

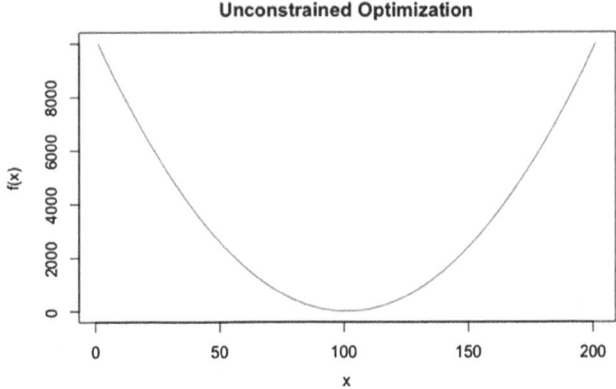

Figure 3-1. *Visualization of f(x)*

In this problem, because there are no constraints, we are allowed to choose whatever number for x is within the bounds defined. Given the equation we seek to minimize, the answer for x is 100. As we can see, we minimize the value of f(x) globally when we choose x. Therefore, we state that x = 100 = x*, which is a global minimizer of f(x). In contrast, here's a constrained optimization problem:

$$\text{Minimize} f(x), \text{ s.t.} (\text{subject to}) x \in \Omega$$

$$\text{where } f(x) = x^2 \text{ Subject to } x \in \Omega$$

The function $f : \mathbb{R}^n \to \mathbb{R}$ that we want to minimize is a real-valued function and is called the objective/cost function. The vector x is a vector of length *n* consisting of independent variables where $x = [x_1, x_2, \ldots, x_n]^T \in \mathbb{R}^n$. The variables within this vector commonly are referred to as *decision variables*. The set Ω is a subset of R called the constraint/feasible set. We say that the preceding optimization problem is a decision problem in which we must find the best vector of x that satisfies the objective subject to the constraint. Here, the best vector of x would result in a minimization of the objective function. In this function, because we have a constraint placed, we call this a constrained optimization problem. $x \in \Omega$ is known as the set constraint. Often, this takes the form of

$$\Omega = \{x : h(x) = 0, g(x) \leq 0\}$$

where *h* and *g* are some given functions. *h* and *g* are referred to as the *functional constraints*.

Imagine that we are still viewing the same function displayed in Figure 3-1, except that our feasible set is Ω. For simplicity's sake, let's say that h(x) and g(x) are equal to the following:

$$h(x) = g(x) = 10 - x$$

As such, the answer for the constrained optimization problem would be x = 10, because this is closest to the global minimizer of f(x), x = 100, while also satisfying the functional constraints listed in Ω. As we can see, the constraint set limits our ability to choose solutions, and therefore compromises must be made. We often encounter constrained optimization in a practical sense on a daily basis. For example, say a business owner is trying to minimize the cost of production in their factory. This would be a constrained optimization problem, due to the fact that should the business owner not want to adversely affect their business (and still continue production), there likely is a production output constraint that they will have placed on them, limiting the possible choices they have.

Most machine learning problems readers will encounter are framed in the scope of a constrained optimization problem, and that constraint is usually a function of the data set being analyzed. The reason for this is often because prior to the development of deep learning models, this was the closest method by which we could approach artificial intelligence. Broadly speaking, most machine learning algorithms focused on regression are constrained optimization problems, where the objective is to minimize the loss of accuracy within a given model. As we briefly discussed in the previous toy problem, there are two kinds of minimizers: local and global.

Local Minimizers

Assume that $f: \mathbb{R}^n \to \mathbb{R}^m$ is a real-values function defined on some set $\Omega \in \mathbb{R}^n$. A point x^* is a *local minimizer* of f over Ω $f(x) \geq f(x^*)$ for all $x \in \Omega$.

Global Minimizers

Assuming the same function *f* and its tertiary properties, a point x* is a global minimizer of *f* over Ω if $f(x) \geq f(x^*)$ for all $x \in \Omega$.

Broadly speaking, there can be multiple local minimizers in a given problem, but if there is a global minimum, there can only been one. In Figure 3-2, we can see this with respect to a mapping of a function.

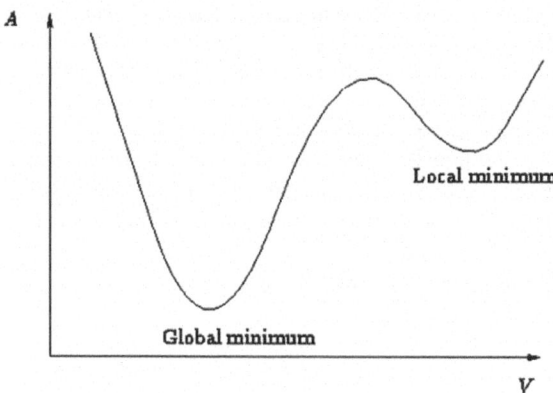

Figure 3-2. *Local versus global minima*

CHAPTER 3 ■ A REVIEW OF OPTIMIZATION AND MACHINE LEARNING

Depending on how much of the function we are evaluating at a given moment, we can choose a multitude of local minima. But if we're evaluating the full range of this function, we can see that there is only one global minimum. It is useful now to discuss how exactly we know that a solution we have reached, mathematically speaking, is optimal.

Conditions for Local Minimizers

In this section, we derive conditions for a point x* to be a local minimizer. We use derivatives of a function $f : \mathbb{R}^n \to \mathbb{R}$. Recall that the first-order derivative of f, denoted Df is

$$Df \triangleq \left[\frac{\partial f}{\partial x_1}, \frac{\partial f}{\partial x_2}, \ldots, \frac{\partial f}{\partial x_n} \right]$$

The gradient of f is just the transpose of Df. The second derivative, or the Hessian of f, is

$$F(x) \triangleq D^2 f(x) = \begin{pmatrix} \frac{\partial^2 f(x)}{\partial x_1^2} & \cdots & \frac{\partial^2 f(x)}{\partial x_n \partial x_1} \\ \vdots & \ddots & \vdots \\ \frac{\partial^2 f(x)}{\partial x_1 x_n} & \cdots & \frac{\partial^2 f(x)}{\partial x_n^2} \end{pmatrix}$$

The first derivate/gradient gives us the direction of an approximation of the function, f, at a specific point. The second derivative, or Hessian, gives us a quadratic approximation of f at a point. Both the Hessian and the gradient can be used to find local solutions for optimization problems. The gradient is used, as discussed earlier, for parameter updating such as in linear regression via gradient descent. However, the Hessian also can be used for parameter updating in the context of deep learning. I talk about this more later, but recurrent neural networks typically are used in the case of modeling data that occurs in sequences such as time series or text segments. Specifically, recurrent neural networks often are difficult to train by a product of certain data sequences having long-term data dependencies. When training other deep learning architectures, we encounter training problems due to the very large number of weights. This creates a large Hessian matrix, virtually making Newton's method defunct.

Hessian-free optimization focuses on minimizing an objective function where instead of computing the Hessian, we compute the matrix-vector product. Provided that the Hessian matrix is positive-definite, we converge to a solution. By solving the following equation, we can effectively use Newton's method on the weight matrix to train a network

$$H_p = \lim_{\epsilon \to 0} \frac{\nabla f(\theta + \epsilon d) - \nabla f(\theta)}{\epsilon}$$

where H_p is the matrix-vector product, θ is some parameter (in this case, weights), and d is a user determined value.

In cases where the Hessian matrix is not positive-definite, convergence upon a solution is not guaranteed and leads to radically different results. We can, however, approximate the Hessian matrix using a Gauss-Newton approximant of the Hessian, whereupon the Hessian equals

$$H = J^T J$$

where J is the Jacobian matrix of the parameter.

This yields a guaranteed positive-definite matrix and therefore validates the assumptions necessary to guarantee convergence. Moving forward from regression, I want to discuss one of the mathematical underpinnings of classification algorithms: neighborhoods.

Neighborhoods

Neighborhoods are an important concept in the paradigm of classification algorithms. For example, the preeminent algorithm that uses this concept is K-nearest neighbors. One of the simpler algorithms, the user-defined K parameter determines the number of neighboring data points that are used to ultimately classify an object to a class of points. We define a *neighborhood* of a point as a set of points containing the aforementioned point without leaving the set. Consider a point $x \in \mathbb{R}^n$. A neighborhood of this set would be the equation

$$\left\{ y \in \mathbb{R}^n : \|y - x\| < \epsilon \right\}$$

where ϵ is some positive number defined in a given context. ϵ represents the bound that defines the size of a given neighborhood. Visually, we can consider a neighborhood as a sphere, or a space between two half spaces, with x as the center and ϵ as the radius, as in Figure 3-3.

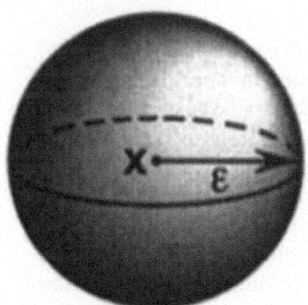

Figure 3-3. *Visualization of a neighborhood of a point x*

This will be an important to understanding any algorithms that use epsilon intensive loss to define separation of observations. Epsilon intensive loss is used particularly in the case of support vector machines, but K-nearest neighbors draws upon the concept of neighborhoods broadly to define observations within a given class.

Interior and Boundary Points

A point $x \in S$ is said to be an interior point of the set S if the set S contains some neighborhood of x. If all points within some neighborhood of x are also in S, the set of all the interior points of S is called the interior of S. A point x is said to be a boundary point of the set S if every neighborhood of x contains a point in S and a point not in S. Similarly, all the boundary points of S take the name boundary. A set is *open* if it contains a neighborhood of each of its points, or has no boundary points. A set is *closed* if it contains its boundary. A set is *compact* if it is both closed and bounded.

We've now reached the conclusion of our review of optimization. Now that we have addressed the prerequisite information necessary, we can discuss machine learning in depth and grasp the broader implications of the algorithms within this paradigm.

Machine Learning Methods: Supervised Learning

Machine learning can be segregated into two broad paradigms: supervised and unsupervised learning. *Supervised learning* is distinguished by the fact that prior to fitting a model, we know what the label/response variable Y is. As such, we can evaluate the efficacy of a model in an efficient manner. In *unsupervised learning*, we don't have this information, which doesn't allow us to determine the degree to which we are correct. Prior to discussing the challenges of both paradigms, it's reasonable to discuss the development of this field

History of Machine Learning

Machine learning was developed to create artificial intelligence in the mid-1950s. Its focus shifted towards creating programs that improved upon iteration, but were specifically made to accomplish one task and could generally be viewed as a method of function optimization. Artificial intelligence eventually began to become its own field, and as the end of the 20th century came, machine learning started to become a more developed and mature science. Machine learning takes contributions and inspiration from many fields, such as statistics and computer science, and the overlap is such that many statistics programs often include and encourage their students to become well versed in the techniques. The upcoming sections will address some of the most common machine learning algorithms, including some of those that serve as inspiration for the deep learning models described in the following chapters.

What Is an Algorithm?

Prior to this point, I've occasionally referred to algorithms. Simply stated, an *algorithm* is a process that we create for the purpose of accomplishing some task. In the following section, prior to tackling deep learning models in the next chapters, we will review important machine learning algorithms that will be utilized in deep learning models in addition to algorithms that are useful in the general practice of data science.

Regression Models

Regression refers to a set of problems in which we are trying to predict specific values. These could be prices of homes, the salary of an employee, or the length of a flower petal. More importantly, regression can also be used to measure the degree to which an explanatory variable(s), x, affect the response variable, Y.

Linear Regression

Imagine we're trying to predict the television ratings of a given show. We know, due to prior research, that the most popular demographic for this show is people aged 25–50 years old. We also see that there is a strong linear correlation between these two variables. As such, we decide to consider this our explanatory, or x, variable and the ratings our response, or Y, variable. How exactly would we proceed? Simple linear regression would be the most logical method. *Simple linear regression* utilizes relatively basic concepts for modeling explanatory variable(s), x, to a response variable, Y. Here we have the model

$$E(Y) = \beta_0 + \beta_1 x_1 + \ldots + \beta_k x_k,$$

where β_0 is the y-intercept and β_1 through β_k are the partial slopes corresponding to each explanatory variable x_1 through x_k, where $k = 1, 2, \ldots, m$, and m = the number of explanatory variables. This is known as a l*inear probabilistic model*, because we're modeling the expectation of Y based on the assumption that it lies somewhere within a distribution of possible points from the ordinary least squares prediction of the point.

Ordinary Least Squares (OLS)

Ordinary least squares is the most basic form of linear regression. The intuition behind why we pick that specific E(Y) value at a specific point x is that we want to find a value for E(Y) that minimizes the squared difference between the actual and predicted Y. When the preceding assumptions are met in a given experiment, we find that the OLS method yields minimum variance and unbiased estimator of Y and also is the maximum likelihood estimator for Y.

Assumptions underlying this model are the following:

- Error terms are normally distributed.
- There is constant variance when observing the error terms.
- Observations of data are independently and identically distributed.
- There is no multicollinearity across explanatory variables.

The intuition behind why we pick that specific E(Y) value at a specific point x is that we want to find a value for E(Y) that minimizes the squared difference between the actual and predicted Y. When the preceding assumptions are met in a given experiment, we find that the OLS method yields minimum variance and unbiased estimator of Y and also is the maximum likelihood estimator for Y. Imagine we have an xy plot, similar to the one show in Figure 3-4.

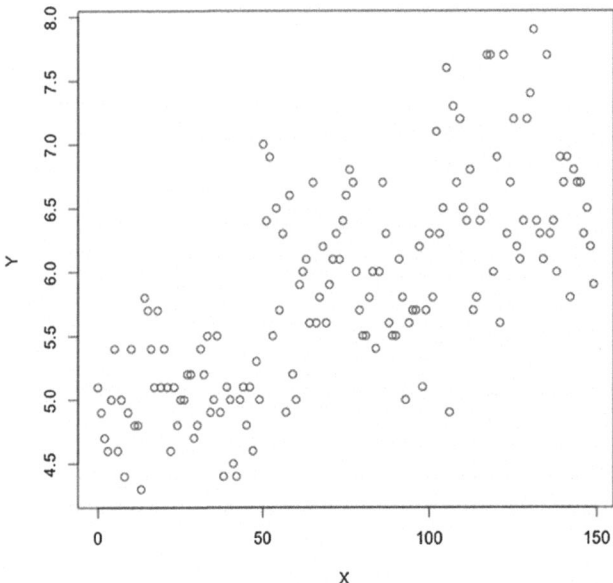

Figure 3-4. *Plotting of the response variable x*

There are in theory an infinite number of E(Y) line plots that we could make. However, only one solution yields an optimal solution that minimizes the error between E(Y) and y the most. Assuming there is only one explanatory variable, we derive the regression coefficient as the following:

$$\hat{\beta} = \left(\frac{\sum_{i=1}^{n} x_i y_i - \frac{1}{n}\left(\sum_{i=1}^{n} x_i\right)\left(\sum_{i=1}^{n} y_i\right)}{\sum_{i=1}^{n} x_i^2 - \frac{1}{n}\left(\sum_{i=1}^{n} x_i\right)^2} \right)$$

Alternatively, the regression coefficient equation can be written

$$\hat{\beta} = \arg\min_{\beta} \|y - x\beta\|$$

given by

$$\hat{\beta} = (X^T X)^{-1} X^T y$$

The purpose of what we are doing here is to minimize the magnitude of the regression coefficient so that when we multiply it by x. Simply stated, we are trying to fine a line of best fit between the data and the regression line such that we minimize the average error between the predictions and actual data points. After we derive the regression coefficient, we can find the y-intercept, or the value of y when x = 0, as the following:

$$\beta_0 = \bar{y} - \bar{x}\beta$$

From this, we have all of the components of the E(Y) equation and can now model the data.

That said, using OLS to find a solution is not always the most optimal method. In cases of relatively small and simple data, utilizing OLS isn't particularly a problem. When data is complex and large, and we haven't satisfied the assumptions of OLS regression, it can be more effective to utilize the gradient descent method.

Gradient Descent Algorithm

As mentioned, the gradient of a function represents the point of greatest rate of increase in the function, and its magnitude is the slope of the graph in that direction. With that in mind, how can we apply the concept of a gradient to an algorithm in order to iteratively improve that? Gradient descent is an iterative algorithm in which you update a parameter by the negative of the gradient subject to some threshold you define or a certain number of iterations. The gradient is usually multiplied by a learning rate, which determines the speed of convergence toward an optimal solution for the function.

In the context of linear regression, our goal is to minimize the residual value between y^ and y, known as the *error function*, given by

$$J(\theta_0, \theta_1) = \frac{1}{2m} \sum_{i=1}^{m} \left(h_\theta(x^i) - y^i \right)^2$$

where $h_\theta(x^i)$ is the predicted y value.

CHAPTER 3 ■ A REVIEW OF OPTIMIZATION AND MACHINE LEARNING

If our objective is to minimize the cost function as quick as possible, and the gradient is the vector that points in the steepest direction, we want to take the gradient of the cost function. The gradient is given by the following:

$$\frac{d}{d\theta_0}(\theta_0, \theta_1) = \frac{1}{m}\sum_{i=1}^{m}\left(h_\theta(x^i) - y^i\right)$$

$$\frac{d}{d\theta_1}(\theta_0, \theta_1) = \frac{1}{m}\sum_{i=1}^{m}\left(h_\theta(x^i) - y^i\right)(x_j)$$

To update the parameters, both the y-intercept and the regression coefficient, we calculate the following until the algorithm converges upon an optimal solution:

$$\theta_0 := \theta_0 - \alpha \frac{d}{d\theta_0}(\theta_0, \theta_1)$$

$$\theta_1 := \theta_1 - \alpha \frac{d}{d\theta_1}(\theta_0, \theta_1)$$

Multiple Linear Regression via Gradient Descent

The intuition behind multiple linear regression via gradient descent is the same as with simple linear regression—there is just a modification to accommodate for the multiple partial slopes being adjusted upon each iteration:

$$\theta_0 := \theta_0 - \alpha \frac{d}{d\theta_0}(\theta_0, \theta_1, \ldots, \theta_n)$$

$$\theta_j := \theta_j - \alpha \frac{d}{d\theta_j}(\theta_0, \theta_1, \ldots, \theta_n)$$

Learning Rates

One last aspect to discuss is the *learning rate,* denoted as α, which in fact is one of the most important aspects of the gradient descent algorithm. The learning rate determines the speed at which the gradient descent algorithm converges upon an optimal solution. Usually, the learning rate is initialized at a relatively small value—typically .01 or less. That said, choosing an optimal learning rate isn't necessarily always obvious, and not doing so can affect the "solution" yielded. Usually, gradient descent algorithms have two stopping conditions: 1) an optimal solution has been found, and 2) the maximum number of iterations allowed have been reached. The following problems associated with poor algorithm performance are due to the following situations:

- *The learning rate is too small*: In the instance that we choose a learning rate that's too small, the solution that the algorithm gives is in fact not the optimal solution, and we reach it due to the second stopping condition. Some might say that a way to avoid this is by choosing a learning rate is to increase the number of iterations, but that very well can defeat the purpose of this method, which is its computational efficiency.

- *The learning rate is too large*: If we are to choose a learning rate that is considerably larger than necessary, we may also never reach an optimal solution, though this one is due to a different reason. When the learning rate is too large, we find that the cost function upon each iteration may overcorrect and give us updated values for the coefficient that are far too small or far too large. As such, our reaching a solution would be by luck, and in most cases we would end up reaching the maximum solution.

Choosing An Appropriate Learning Rate

Now that we have an understanding of what the problems associated with choosing an incorrect learning rate are, we need to find out how to choose one. One possible solution is to hardcode various gradients and see how the algorithms perform across each iteration. In the following method, we update the step size upon each iteration of the gradient descent algorithm.

The bold driver approach compares the most recent gradient value to the gradient value derived upon the prior iteration. If the error has decreased, increase the learning rate by a moderate amount. If the error has increased, decrease the learning rate by 50%.

In the following code example, we are modifying the iris data set. This data set dates back to Ronald Fisher; he used it for an initial set of experiments. It is popular when displaying fundamental aspects of various statistical and machine learning algorithms. Here, we're taking the first column of the iris data set and modeling that against an X variable (which is merely length), such that the data displayed forms a linear pattern. This is just an example to display the mechanics of OLS linear regression. In the following code, we fit the data to the OLS regression via the lm( ) function. We then calculate the sum of squared residuals, which we denote as Cost in the output. We then extract the regression coefficients for this model from the lm( ) function and then output these two attributes in a data frame:

```
#Modifying Data From Iris Data Set
data(iris)
Y <-  matrix(iris[,1])
X <-  matrix(seq(0,149, 1))
```

CHAPTER 3 ■ A REVIEW OF OPTIMIZATION AND MACHINE LEARNING

```
olsExample <- function(y = Y, x = X){
  y_h <- lm(y ~ x)(1)
  y_hf <- y_h$fitted.values
  error <- sum((y_hf - y)^2) (2)
  coefs <- y_h$coefficients (3)
  output <- list("Cost" = error, "Coefficients" = coefs)
  return(output)
}
```

When we run the code, we observe the results shown in Figure 3-5.

$Cost
[1] 49.69214

$Coefficients
(Intercept)   x
4.82567770   0.01365981

Figure 3-5. *Output of OLS regression function*

Cost is the sum of squares and the Coefficients accordingly are listed as the y intercept followed by the partial slope for the x variable. We will use this as a baseline for comparing the performance of linear regression via gradient descent. Again, the purpose of this explanation is to show the efficiency and ability of the gradient descent algorithm to replicate the results of a simple OLS regression:

```
#Gradient Descent Without Adaptive Step
gradientDescent <- function(y = Y, x = X, alpha = .0001, epsilon = .000001,
maxiter = 300000){
    #Intializing Parameters
    theta0 <-  0
    theta1 <-  0
    cost <- sum(((theta0 + theta1*x) - y)^2)
    converged <- FALSE
    iterations <- 1
```

Moving forward, we define a function for the implementation of linear regression via gradient descent. This gradient descent algorithm has a constant learning rate, though you can alter this parameter, as well as the loss tolerance, should you choose to use this implementation on other data sets. We have defined the maximum amount of iterations as 300,000, which will force the algorithm to cease at the solution should it not reach an optimal one before that. When analyzing the code specifically, we begin by initializing the parameters theta0 and theta1 at 0. Users may feel free to alter the code and initialize the parameters with values randomly sampled from a normal distribution, but should divide these values by 10 to ensure that they are not overly large. We initialize the cost function as the SSR of 0 minus all y values, from which we will begin to alter the parameters:

```
#Gradient Descent Algorithm
while (converged == FALSE){
  gradient0 <- as.numeric((1/length(y))*sum((theta0 + theta1*x) - y))
  gradient1 <- as.numeric((1/length(y))*sum((((theta0 + theta1*x) - y)*x)))

  t0 <- as.numeric(theta0 - (alpha*gradient0))
  t1 <- as.numeric(theta1 - (alpha*gradient1))

  theta0 <- t0
  theta1 <- t1

  error <- as.numeric(sum(((theta0 + theta1*x) - y)^2))

  if (as.numeric(abs(cost - error)) <= epsilon){
    converged <- TRUE
  }
    cost <- error
    iterations <- iterations + 1
  if (iterations == maxiter){
    converged <- TRUE
  }
}
```

Although we haven't converged on a solution, or we have not reached the maximum amount of iterations allowed when executing the function, we create the gradient0 and gradient1 variables, which correspond to the parameters theta0 and theta1 respectively. We then update the theta0 and theta1 parameters using the gradient contained within the gradient0 and gradient1 variables. After this, we calculate the error, and continue looping from while (converged == FALSE) until the stopping condition has been reached:

```
  output <- list("theta0" = theta0, "theta1" = theta1, "Cost" = cost,
  "Iterations" = iterations)
  return(output)
}
```

Here, we're running a simple linear regression where the y and x variables are initialized randomly. When we run the code as stated, we get the results shown in Figure 3-6.

```
$theta0
[1] 4.102621

$theta1
[1] 0.0209148

$Cost
[1] 69.49426

$Iterations
[1] 75177
```

Figure 3-6. *Output of gradient descent without adaptive step function*

theta0 is the y-intercept, theta1 is the partial slope for the x variable, cost is the sum of squares, and iterations is the number of iterations performed. Here, we observe lower regression coefficients that are roughly the same baseline sum of squares error. However, if we're to use a learning rate that's too large, we often will get an error because the regression coefficients have become infinitely large. When the learning rate is too small, we notice what's shown in Figure 3-7.

```
$theta0
[1] 2.539109

$theta1
[1] 0.03660275

$Cost
[1] 247.7242

$Iterations
[1] 295771
```

Figure 3-7. *Output of gradient descent with small learning rate*

We see that the algorithm doesn't converge upon the minimum, but reaches a feasible solution and is cut off by the loss tolerance we set. Incidentally, we're also near the maximum number of iterations allowed. The consequence of incorrectly choosing an algorithm is relative to the context in which a given algorithm is being applied. But all

CHAPTER 3 ■ A REVIEW OF OPTIMIZATION AND MACHINE LEARNING

users should be careful to evaluate the results they find on any machine learning or deep learning algorithm. As such, it's important that we are as confident as humanly possible when choosing a solution.

In the next example, we run the same algorithm using an adaptive step size to compare the performance:

```
#Gradient Descent with Adaptive Step
adaptiveGradient <- function(y = Y, x = X, alpha = .0001, epsilon = .000001,
maxiter = 300000){
  #Intializing Parameters
  theta0 <-  0
  theta1 <-  0
  cost <- sum(((theta0 + theta1*x) - y)^2)
  converged <- FALSE
  iterations <- 1
  
  #Gradient Descent Algorithm
  while (converged == FALSE){
    gradient0 <- as.numeric((1/length(y))*sum((theta0 + theta1*x) - y))
    gradient1 <- as.numeric((1/length(y))*sum((((theta0 + theta1*x) - y)*x)))
    
    t0 <- as.numeric(theta0 - (alpha*gradient0))
    t1 <- as.numeric(theta1 - (alpha*gradient1))
    
    delta_0 <- t0 - theta0
    delta_1  <- t1 - theta1
    if (delta_0 < theta0){
      alpha <- alpha*1.10
    } else {
      alpha <- alpha*.50
    }
```

Here, we apply the same gradient descent function, except now we apply the bold driver approach so that we have an adaptive learning rate. The bold driver approach alters the learning rate from one individual iteration to the next based on the prior result. Simply stated, if the gradient increases from one iteration to the next, the learning rate increases by 10%. If the gradient decreases, we decrease the learning rate by 50%. Readers can feel free to alter these parameters should they choose, to experiment on the results received:

```
    theta0 <- t0
    theta1 <- t1
    error <- as.numeric(sum(((theta0 + theta1*x) - y)^2))
    if (as.numeric(abs(cost - error)) <= epsilon){
      converged <- TRUE
    }
    cost <- error
    iterations <- iterations + 1
    if (iterations == maxiter){
```

```
            converged <- TRUE
        }
    }
    output <- list("theta0" = theta0, "theta1" = theta1, "Cost" = cost,
    "Iterations" = iterations, "Learning.Rate" = alpha)
    return(output)
}
```

Upon executing this function, we receive the results shown in Figure 3-8.

```
$theta0
[1] 4.667928

$theta1
[1] 0.01527666

$Cost
[1] 50.63598

$Iterations
[1] 34352

$Learning.Rate
[1] 0.0003044947
```

Figure 3-8. Output of gradient descent with adaptive learning rate functions

Of the algorithms we have tested, and given our objective to minimize the cost and the regression coefficient size, linear regression via gradient descent *with* adaptive step size or the OLS method would be acceptable. As an interesting observation, the algorithm converged upon this solution significantly faster than the gradient descent method with a static learning rate did.

Newton's Method

For instances in which we're looking to minimize a quadratic function, Newton's method often proves useful. Newton's method is a way to find the roots of a function, or where f(x) is equal to 0. It was developed by Isaac Newton and Joseph Raphson. To calculate an optimal point, we derive the equation

$$x^{k+1} = x^k - \left(\frac{f'(x^k)}{f''(x^k)}\right)$$

where f' is the first derivative of a given function and f" is the second derivative of a given function. This is known as the *secant method*. Newton's method works particularly well if the f'"(x) > 0, but if f"(x) < 0, it might not converge upon a global minimum. We know that Newton's function will always converge to a global optimum if the Hessian of the function is positive semi-definite. Another drawback to Newton's method is also that convergence is not guaranteed if the starting point is considerably far from the global minimum. In the instance that Newton's method doesn't converge upon the global minimum, there is a heuristic that can be used to overcome this, covered next.

Levenberg-Marquardt Heuristic

The Levenberg-Marquardt (LM) algorithm is most applicable when a function isn't twice differentiable or its Hessian matrix isn't positive-definite. The equation is given by the following:

$$x^{k+1} = x^k - \left(F(x^k) + \mu_k I\right)^{-1} g^k$$

Consider a square matrix F that isn't positive-definite. The eigenvalues of this matrix may not be positive but are all real numbers. Consider a matrix

$$G = F + \mu I$$

where $\mu \geq 0$. The eigenvectos of G are $\lambda + \mu$. Therefore, the following must be true:

$$Gv_i = (F + \mu I)v_i$$

$$= Fv_i + \mu I v_i$$

$$= \lambda_i v_i + \mu v_i$$

$$= (\lambda_i + \mu)v_i,$$

With this modification, all the eigenvalues of G are therefore positive, and then G would have to be positive definite. If μ is also sufficiently large enough, we can confirm that the direction that Newton's algorithm chooses will always be toward the direction of steepest descent. The final modification to the algorithm will be to add in a step size:

$$x^{k+1} = x^k - \alpha \left(F(x^k) + \mu_k I\right)^{-1} g^k$$

What Is Multicollinearity?

Multicollinearity is a problem that many a data scientist will come across in the problems they solve. It's a situation where the explanatory variables are nearly perfectly correlated with other. In this situation, it becomes difficult to use linear regression via OLS or gradient descent because the technique cannot accurately estimate the regression coefficients, often resulting in inflated values for these parameters. This is because it's hard to distinguish the effect of one explanatory variable from another, and subsequently each explanatory variable's effect on the response variable. As a product of multicollinearity, we observe the value of the regression coefficients changing, sometimes drastically, each time we initialize a linear regression algorithm. Ultimately, this renders traditional linear regression as a less preferable method for handling data that exhibits these types of patterns.

Testing for Multicollinearity

Very highly positive regression coefficients are one of the first tell-tale signs of multicollinearity. In addition to this, we should calculate the correlation of all the explanatory variables with each other. Correlation coefficients of $.95 \leq \rho \leq 1$ should also raise red flags in the mind of a data scientist. Specifically, though, there is a statistic that we can use to determine whether we most definitely have multicollinearity in our data set, called variance inflation factor.

Variance Inflation Factor (VIF)

The VIF statistic is calculated on a range from $0 \leq VIF \leq \infty$. Typically, the rule of thumb is that any VIF score that is > 5 indicates multicollinearity, and any score above 10 indicates severe multicollinearity. The statistic is calculated by regressing a given explanatory variable against the others and then using the result to calculate the coefficient of determination, yielding the following:

$$VIF_j = \frac{1}{1-R_j^2}, \text{ Where } j = 1,\ldots,k$$

Ridge Regression

To combat multicollinearity specifically, ridge regression was developed and is a useful technique. Relevant to our discussions of norms earlier (L1 versus L2), ridge regression uses an L2 norm to achieve an optimal solution. Here is the equation for ridge regression:

$$\arg\min_{\beta} \|y - X\beta\|_2^2 + \lambda \|\beta\|_2$$

One of the key distinctions in ridge regression is the tuning parameter λ, which determines the degree to which the regression coefficients shrink. The technique gets the name *ridge* due to the fact that the L2 norm forms a spherical or circular shaped region where the optimal solutions for the regression coefficients exist are chosen along the "ridge" of this shape. Visually, this often looks like Figure 3-9.

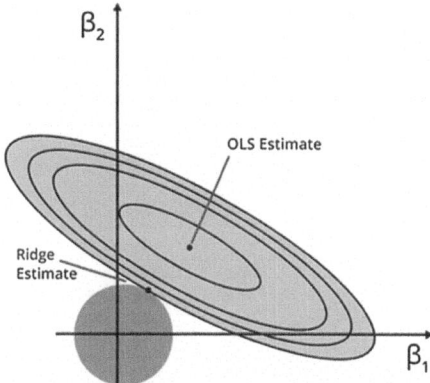

Figure 3-9. *Ridge regression OLS estimates*

Least Absolute Shrinkage and Selection Operator (LASSO)

Lasso is very similar to ridge regression except LASSO performs variable selection while regressing the explanatory and response variables. The key differentiation between LASSO and ridge regression is the fact that LASSO uses the L1 norm rather than the L2 norm, giving the selection region a square or cubic shape depending on the dimensionality of the data. In Figure 3-10, we can see the LASSO OLS estimate:

$$\arg\min_{\beta} \|y - X\beta\| + \lambda \|\beta\|$$

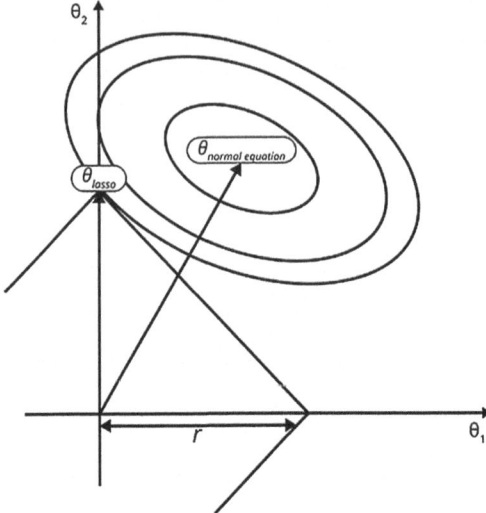

Figure 3-10. LASSO regression

Comparing Ridge Regression and LASSO

Both methods are highly useful for instances in which your data suffers from multicollinearity, but in instances where you're seeking to fit data as you would in a simple linear regression, these methods should be avoided, and the gradient methods along with the OLS method given earlier should be used. If you don't have more than one explanatory variable, these methods won't be of use to you. Although that's unlikely to be the case in practical terms most times, it's important to remember nonetheless.

Evaluating Regression Models

Beyond just building regression models, we need to find a way to determine how accurately the results yielded from a model are, and ultimately choose the best one on a case-by-case basis. In the case of regression, a useful method of evaluating machine learning models is by bootstrapping. Typically, *bootstrapping* involves running different regression models over several iterations using a data set that's smaller than the original, and with the original observations in randomized order, and then sampling several statistics and comparing their values relative to the other models' values. The process is as follows:

1. Build several models.

2. Collect sample statistics that we use as evaluators of each model over N iterations of the experiment.

CHAPTER 3 ■ A REVIEW OF OPTIMIZATION AND MACHINE LEARNING

3. Sample each of these evaluators and collect statistics upon each iteration, such as:

   a. Mean

   b. Standard deviation

   c. Max

   d. Min

4. Evaluate the results and pick the model that's most effective given your objectives and situational constraints.

The rest of this section covers the evaluators you should pick during bootstrapping.

Coefficient of Determination (R^2)

As described in Chapter 2, the coefficient of determination is what we use to evaluate how accurately a model explains variability in y through the variability in x. The higher the R^2 value, the better. That said, generally speaking, "good" R^2 values should be in the following range: $.70 \leq R^2 \leq .95$. Anything lower than .70 should be viewed as generally unacceptable, and anything higher than .95 should be examined to see if there is overfitting in the model. Although this won't change across a given iteration very much, we still should evaluate this objectively across models.

Mean Squared Error (MSE)

The MSE measures the distance of a given predicted value of y from the average value of the actual response variable. Our objective with any regression model is to minimize this statistic as much as possible, so we will want to pick the model that has the lowest MSE relative to the others being examined. This will be the evaluator that shows the most variance across models and should be the one that gives us the most inferential power with respect to which model we should choose.

Standard Error (SE)

In the case of a regression model, we would probably measure the standard error of a given model. The objective we should have should be to have a standard error that is as close to 0 as possible. Highly negative or highly positive standard error values are generally undesirable.

Classification

Moving beyond the case of predicting specific values, our data observations often belong to some class that we would like to label them as such. We refer to this paradigm of problems as *classification problems*. To introduce readers to these types of problems, we begin by addressing the most elementary of these algorithms: logistic regression.

Logistic Regression

In addition to regression, one of the important tasks of machine learning is classification of an observation. Although there are multinomial classification algorithms, we will start by examining a *binary classifier*, a method often used as a baseline for the remainder of the classifiers. Logistic regression gets its name from the function that powers it, known as the *logistic function*, illustrated in Figure 3-11.

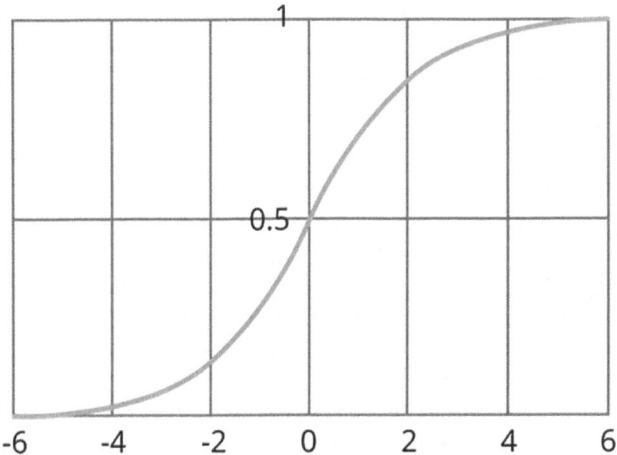

Figure 3-11. Visualization of logistic function

The function itself reads this this:

$$f(x) = \left(\frac{1}{1+e^{-x}}\right)$$

The intuition behind how we classify an observation is simple: we set a threshold for a given $f(x)$ value and then classify it as a 1 if it meets or exceeds this threshold and a 0 if otherwise. In many contexts, the x variable will be replaced by a linear regression formula, in which we model the data. As such, the equation for $f(x)$, or the log odds, will be

$$\pi = \frac{1}{1+e^{-(\beta_0+\beta_1 X)}}$$

where π = log odds and π^* is the given threshold. As for the threshold we establish, that depends on what we would like to maximize: accuracy, sensitivity, or specificity.

- *Sensitivity/recall*: The ability of a binary classifier to detect true positives:

$$True\ Positive\ Rate = \frac{True\ Positives}{Positives} = \frac{True\ Positives}{True\ Positive + False\ Negatives}$$

- *Specificity*: The ability of a binary classifier to detect true negatives:

$$Specificity = \frac{True\ Negative}{Negatives} = \frac{True\ Negatives}{True\ Negatives + False\ Positives}$$

- *Accuracy*: The ability of a binary classifier to accurately classify both positives and negatives:

$$Accuracy = \frac{True\ Positives + True\ Negatives}{Positives + Negatives}$$

In certain contexts, it may be more advantageous to magnify any of these statistics, but that's all relative to where these algorithms are being applied. For example, if you were testing the probability of a phone battery combusting, you would probably want to be certain that false negatives are minimized as much as possible. But if you were trying to detect the probability that someone is going to find a match on a dating website, you probably would want to maximize true positives. The relationship between the tradeoff of these predictive abilities is most easily exemplified using an ROC curve, which shows how altering the value of π^* affects the classification statistics of the model.

Receiver Operating Characteristic (ROC) Curve

The ROC curve initially was used during World War II for the purposes of radar detection, but its uses were soon considered for other fields, statistics being one of them. The ROC curve displays the ability of a binary classifier to accurately detect true positives and simultaneously check how inaccurate it is by displaying its false positive rate. This is shown in Figure 3-12.

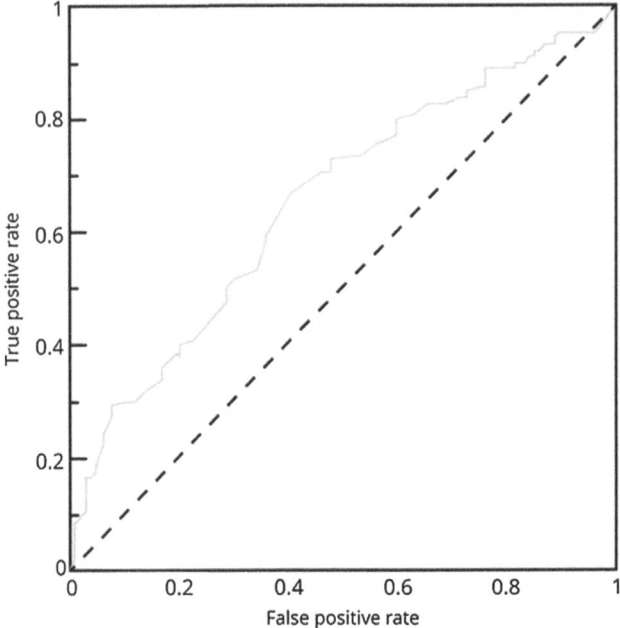

Figure 3-12. Example ROC curve plot

In the context of logistic regression, the evaluation of any specific model, given a specific threshold for π, is ultimately determined by the *area under the curve*, or AUC. The vertex of the plot is the .50 AUC score, which indicates that the model, should its score be this, is no better at classifying than a random guess. Ideally, this AUC score would be as close to one, but we generally accept anything ≥ .70 as acceptable.

Confusion Matrix

Another method of evaluating classification models is the *confusion matrix*, a graphical representation of the classifiers predictions against the actual labels of a given observations. From this visualization, we derive the values for the statistics listed previously that ultimately help us accurately evaluate a classification model's performance. Figure 3-13 shows a visual example of a confusion matrix.

CHAPTER 3 ■ A REVIEW OF OPTIMIZATION AND MACHINE LEARNING

	P' (Predicted)	n' (Predicted)
P (Actual)	True Positive	False Negative
n (Actual)	False Positive	True Negative

Figure 3-13. Confusion matrix

Interpreting the values within a confusion matrix often is a subjective task that is up to the reader to determine. In some instances, false positives, such as determining whether users should buy a product or not, will not be as detrimental to solving the problem at hand. In other cases, such as determining whether a car engine is faulty, false positives may actually be detrimental. Readers should be conscious of the task they are performing and tune the model to limit the false positives and/or false negatives accordingly.

Limitations to Logistic Regression

Logistic regression can only predict discrete outcomes. It requires many of the assumptions necessary for ordinary linear regression, and overfitting of data can become quite common. In addition to this, classification with logistic regression works best when we have data that is clearly separable. For these reasons, in addition to the fact that there are more sophisticated techniques available, it is a common modeling practice to consider the logistic regression model to be the baseline by which we juxtapose other classification methods and observe the nuances.

Moving forward, we will look at a simple example using logistic regression. This data set will be referenced in later chapters, and in Chapter 10 in detail, for those who are curious about the process by which this model was produced.:

```
#Code Redacted, please check github!
#Logistic Regression Model
lr1 <- glm(data[,1] ~ data[,2] + data[,3] + data[,4] + data[,5] + data[,6]
+ data[,7],
           family = binomial(link = "logit"), data = data)

#Building Random Threshold
y_h <- ifelse(lr1$fitted.values >= .40, 1, 0)
```

CHAPTER 3 ■ A REVIEW OF OPTIMIZATION AND MACHINE LEARNING

```
#Construct ROC Curve
roc(response = data[,1], predictor = y_h, plot=TRUE, las=TRUE,
    legacy.axes=TRUE, lwd=5,
    main="ROC for Speed Dating Analysis", cex.main=1.6, cex.axis=1.3,
    cex.lab=1.3)
```

The ROC curve for our model is shown in Figure 3-14.

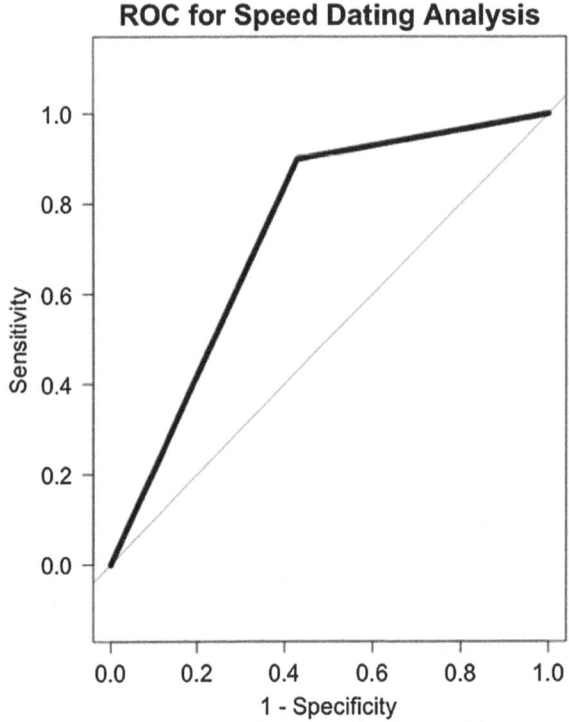

Figure 3-14. ROC curve for logistic regression example

Using the preceding code, we have an area under the curve of .7353. Given the threshold that we set before, this model's performance would be considered acceptable, but it should likely undergo more tuning.

Support Vector Machine (SVM)

Among the more sophisticated machine learning models available, *support vector machines* are a binary classification method that has more flexibility than the logistic regression model in that they can perform non-linear classification. This is performed via its kernel functions, which are equations that orthogonally project the data onto a new feature space, and the classification of the objects are performed as a product of two hyperplanes constructed by a norm (See Chapter 2).

In the case of liner SVMs, we take in as our inputs a response variable, Y, and an explanatory variable(s), x. We orthogonally project this data into feature space, such that we form the hyperplane that separates the data points. The size of these hyperplanes is determined by the Euclidean norm of the weights, or w, vector in addition to an upper bound and lower bound, respectively denoted as the following:

$$wx + b = 1$$

$$wx + b = -1$$

We keep reiterating this process until we have reached a norm of w that maximizes the separation between the two classes. The separation of the classes is maximized by minimizing ||w||, being that the size of the hyperplane is given by the following:

$$\frac{2}{\|w\|}$$

The following constraints also prevent us from allowing observations to fall in between the two hyperplanes:

$$wx + b \geq 1, \text{ if } y = 1$$

$$wx + b \leq -1, \text{ if } y = -1$$

The observations that ultimately fall on the boundaries of the hyperplane are the most important, as they are the "support vectors" that define the separation between classifications. This transformation is shown in Figure 3-15.

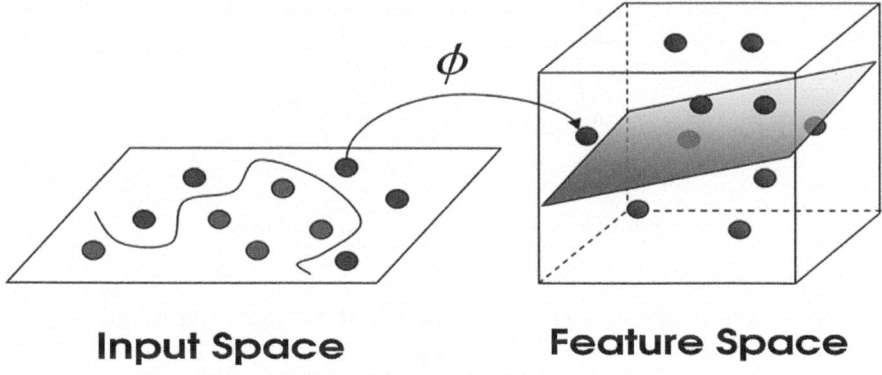

Figure 3-15. *Orthogonal transformation of data via kernel function*

The optimization problem formed from these constraints in addition to objective of the SVM is given by the following:

$$\text{Minimize} \|w\| \text{ subject to } y_i(wx+b) \geq 1, \text{ for } i=1,\ldots,n$$

Types of Kernels

To expand the flexibility of SVMs, different kernels have been developed. Among them are the following:

- Polynomial
- Gaussian radial basis function
- Hyperbolic tangent

Sub-Gradient Method Applied to SVMs

A sub-gradient of a function is defined as a generalization of a derivative to a function that are not differentiable. Simply stated, it is the slope of a line that goes through the derivative of a function, but falls below the derivative. Modern modifications to the SVM algorithms have yielded better performing classification models when dealing with data with more than 10^5 features and 10^5 observations. We define the optimization problem as

$$f(w,b) = \left[\frac{1}{n}\sum_{i=1}^{n}\max(0, 1 - y_i(wx+b))\right] + \lambda\|w\|^2$$

where f is a convex function of w and b. Moreover, this allows us to use gradient decent methods because they work particularly well on convex sets. Given a cost function C(w), defined as the actual classification minus the predicted classification, we use the gradient descent formula therefore as follows:

$$\min_{w \in \mathbb{R}^d} C \sum_{i=1}^{n} \max(0, 1 - y_i f(x_i)) + \|w\|^2$$

$$\min_{w} C(w) = \frac{\lambda}{2}\|w\|^2 + \frac{1}{n}\sum_{i=1}^{n}\max(0, 1 - y_i f(x_i))$$

The sub-gradient step size selection method is similar to the bold driver approach described earlier in this chapter. As always, the context in which an algorithm is being applied should ultimately decide which method is used, not just which performs better.

Extensions of Support Vector Machines

A regression method proposed in 1996 by Vladimir Vapnik, Harris Druck, Christopher Burges, Linda Kaugman, and Alexander Smola is among the more popular extensions of SVMs. The difference is that in SVR, we don't care about the observations that fall within the hyperplane. Instead, we only modify the shape of the hyperplane in response to points that fall outside the loss tolerance zone, with the objective of minimizing the amount of these points that fall out of this zone. The type of regression performed can be altered again by the same kernel functions listed earlier. In addition to this, an alternative to K-means clustering which uses the Gaussian kernel as the activation function for orthogonal projection. Here, the algorithm searches to make the hyperplane such that the smallest sphere that encloses the image of the data defines a given cluster. Clustering algorithms are covered later in this chapter.

Limitations Associated with SVMs

The main problem with SVMs arises from what is arguably the key to why they are so powerful: kernel functions. Determining the proper kernel to use is often stated as the largest drawback to this technique. When delving deeper into this aspect of algorithm training, specifically the selection of the loss parameter and the Gaussian kernel's width parameter is not apparently obvious and is highly subject to the context in which the algorithm is being used. Second, although SVMs do perform well on large data sets, they are a computationally expensive method and require sufficiently good hardware when applied in an industry setting. As such, it does not always make sense to use SVMs in contexts outside of research, or any context where real-time data would be analyzed.

The following is a quick example of SVMs used on the iris data set:

```
#Code Redacted, please check github!
require(LiblineaR)
require(e1071)

#SVM Classification
output    <- LiblineaR(data=s, target=y_train, type = 3, cost = heuristicC(s))

#Predicted Y Values
y_h <- predict(output, s, decisionValues = TRUE)$predictions

#Confusion Matrix
confusion_matrix <- table(y_h, y_train)
print(confusion_matrix)
```

When executing our code, it yields the confusion matrix shown in Figure 3-16.

```
       y_train
y_h   1   2   3
  1  34   0   0
  2   1  18   8
  3   0  15  24
```

Figure 3-16. Confusion matrix for support vector machine

Machine Learning Methods: Unsupervised Learning

Moving beyond the paradigm in which we know the answers we're trying to predict is the more ambiguous section of deep learning, in which we are trying to make inferences based off of our algorithms. This specific subset of problems is known as being a part of *unsupervised learning*, or problems where we don't know a priori what the answers should be.

K-Means Clustering

Until now, we've spoken primarily about supervised learning, but another important aspect of machine learning is the use of algorithms in unsupervised learning cases. Typically, unsupervised learning can be performed as an exploratory research method, or as a preliminary step prior to the primary component of the experiment. One of the best examples of unsupervised learning is the K-means clustering algorithm. The motivation behind this algorithm is to find observations that are similar based on the distance they are away a cluster center.

Assignment Step

Here, we take the observations of data and give an initial set of k means by calculating the means of three random observations within the data. From this point, we assign each observation to the cluster centers based on which assignment yields the smallest within cluster sum of squares, determined by the Euclidean norm between the observation's mean and the cluster center mean

$$S_i^t = \left\{ x_p : \left\| x_p - m_i^t \right\|^2 \leq \left\| x_p - m_j^t \right\|^2 \ \forall j, 1 \leq j \leq k \right\}$$

where S = cluster center and x_p is an observations mean.

Update Step

We then recalculate the cluster mean by taking the mean of the observations within the center and then reiterate over these two steps until reassignments stop:

$$m_i^{t+1} = \frac{1}{|S_i^t|} \sum_{x_j \in S_i^t} x_j$$

Limitations of K-Means Clustering

The major problem with K-means clustering is that the solution reached is often dependent on where the means are initialized, and therefore convergence upon a global minimum isn't guaranteed. Also, depending on the variation of K-means chosen, the time taken until convergence may also not be particularly fast.

Here's a brief example of K-means clustering:

```
#Upload data
data  <- read.table("http://statweb.stanford.edu/~tibs/ElemStatLearn/
datasets/nci.data", sep ="", header = FALSE)
data <- t(data)
k_means   <- c()
k   <-  seq(2, 10, 1)
for (i in k){
  k_means[i]  <- kmeans(data, i, iter.max = 100, nstart = i)$tot.withinss
}

clus <- kmeans(data, 10)$cluster
summ   <- table(clus)
#Removing NA Values
k_means   <- k_means[!is.na(k_means)]
#Plotting Sum of Squares over K
plot(k, k_means,  main ="Sum of Squares Over K-Clusters", xlab = "K
Clusters", ylab= "Sum of Squares",
     type = "b", col = "red")
```

Typically when performing K-means clustering, the most difficult part is determining which value of k we should pick. Typically, the more clusters one has, the lower the sum of squares within a cluster between its observations and the cluster centers will be. However, the more clusters that are present, the less informative these clusters are. Therefore, the challenge becomes a tradeoff between sum of squares over the K clusters and as least clusters as possible to make the observations reasonably differentiable. Figure 3-17 shows a plot that aids us in that effort.

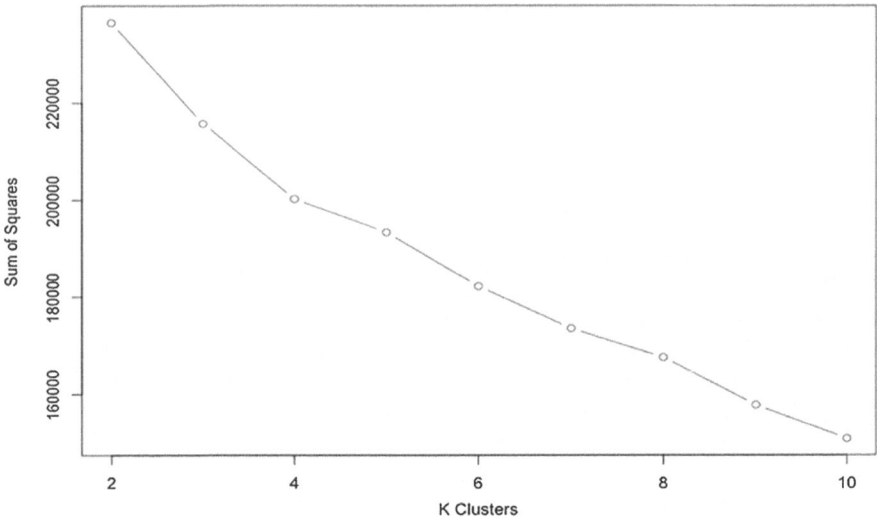

Figure 3-17. *Within cluster sum of squares over K clusters*

In the plot in Figure 3-17, we notice that our sum of square decreases dramatically in the beginning, but we see that a tapering off toward the end in the value changes. As such, it's reasonable for us to choose a value between 6 and 8, preferably closer to, if not, 6. This follows the objectives laid out in the prior paragraph and would yield us with actionable insights, or create a feature for a data sight that contains significant differences for a classification or regression algorithm to detect.

Expectation Maximization (EM) Algorithm

Popular within the paradigm of unsupervised learning, EM algorithms can be used for a multitude of purposes such as classification or regression. Most specifically of use to the user, it can be used to impute values that are missing within a data set. We will show this capability in Chapter 11. Regardless, the EM algorithm is a probabilistic model, which distinguishes it from many machine learning models, which often tend to be deterministic. The algorithm uses the log-likelihood function to estimate the parameter and then maximizes the expected log-likelihood found.

Expectation Step

Consider a set of unknown values Z, which is a subset of the data set X. We calculate the log-likelihood of a parameter with regard to the conditional distribution of Z given X. The following equation yields the expected value of the maximum likelihood estimate of the parameter:

$$L(\theta;X) = p(X|\theta) = \sum_z p(X,Z|\theta)$$

$$Q(\theta|\theta^t) = E_{(Z|X,\theta^t)}\left[\log L(\theta;X,Z)\right]$$

Maximization Step

In this step, we seek to maximize the probability of the given parameter we are analyzing. The equation is given by the following:

$$\theta^{t+1} = \arg\max_{\theta} Q(\theta|\theta^t)$$

Limitations to Expectation Maximization Algorithm

The EM algorithm also tends to be very slow to converge and doesn't yield the asymptotic variance-covariance matrix of the MLE. In addition to this—similar to the same limitation with naïve Bayes classifiers, because the MLE estimator assumes feature independence—it would be ill advised to use this method if the features being analyzed are in fact not independent. The following is an example of the EM algorithm used for classification via clustering:

```
#Expectation-Maximization Algorithm for Clustering
require(MASS)
require(mclust)
y_h <- Mclust(x_train, G = 3)$classification
print(table(y_h, y_train))
plot(Mclust(x_train, G = 3), what = c("classification"), dimens=c(1,3))
```

When executing our code, it yields the plot shown in Figure 3-18.

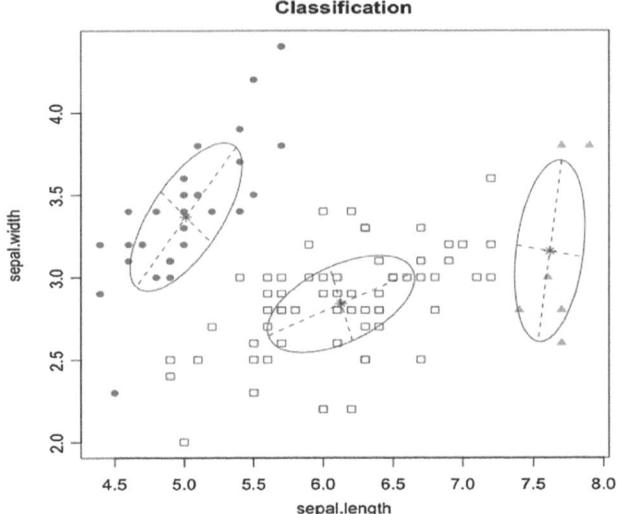

Figure 3-18. *Iris data clusters from EM algorithm clustering*

Decision Tree Learning

Commonly used across a variety of fields for the purpose of data mining, *decision trees* yield a relatively simple method of uncovering insights hidden below the surface of data. There are broadly two types of decision tress, and they typically are used for regression and classification. Decision trees are constructed by creating a rule that determines which direction the decision flows. The idea is that you use a funnel methodology in which the first rule is the broadest and you break down the questions into subsets until the final "leaf" is the most granular aspect determined.

The benefits often associated with decision trees is that overall, they are relatively easy to understand and generally quite effective. In addition, decision trees can handle missing data better than some machine algorithms can without replacing or changing the data (we can just average the values or classifications), and they are quick to compute final values relative to other modeling techniques. Above all, there are varieties of methods that can be used to help the trees learn effectively, and they can model data well when traditional regression methods cannot.

Classification Trees

Classification trees are similar to regression trees. The splits are usually determined by binary variables, but they can be both numerical and categorical. In addition to this, classification trees can make two types of predictions: 1) *point prediction*, which simply denotes the class, and 2) *distributional prediction*, which gives a probability for each class. For probability forecasts, each terminal node in the tree yields a distribution over the classes. If the leaf corresponds to the sequence of answers, given by $A = a$, $B = b$, ... $Q = q$, then the following equation yields the probability:

$$\Pr(Y = y | A = a, B = b, \ldots, Q = q)$$

To evaluate the classification tree, the same methods of evaluating different classification models as described earlier are used. But we also introduce the concept of average loss. Simply stated, some errors are likely to cause greater "damage" toward accurately reaching the correct classification. The *average loss* formula is given by the following:

$$\text{Loss}(Y = j | X = x) = \sum_i L_{ij} \Pr(Y = i | X = x)$$

Moving beyond this, we can determine whether the model made an incorrect classification in cases where it was or was not uncertain using the normalized negative log-likelihood. The formula for it is given by

$$L(data, Q) = -\frac{1}{n} \sum_i^n \log Q(Y = y_i | X = x_i)$$

where $Q(Y = y | X = x)$ is the conditional probability the model predicts. In this context, L is also referred to as *cross-entropy*. If perfect classification were possible, L would be 0. If there is some irreducible uncertainty in the classification, the best possible classifier would give $L = H[Y|X]$, the conditional entropy of Y given X. Less than ideal predictors have $L > H[Y|X]$. Here is an example of a classification tree:

```
require(rpart)
#Classification Tree
classification_tree <- rpart(y_train ~ x_train[,1] + x_train[,2] + x_train[,3] + x_train[,4]
                             +x_train[,5] + x_train[,6], method = "class")
pruned_tree <- prune(classification_tree, cp = .01)

#Data Plot
plot(pruned_tree, uniform = TRUE, branch  = .7, margin = .1, cex = .08)
text(pruned_tree, all = TRUE, use.n = TRUE)
```

When executing our code, we yield the plot shown in Figure 3-19.

CHAPTER 3 ■ A REVIEW OF OPTIMIZATION AND MACHINE LEARNING

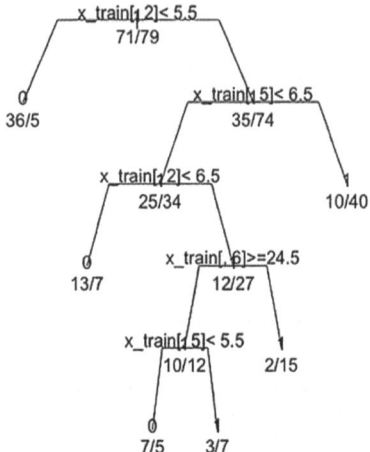

Figure 3-19. Classification tree splits based on Classification tree model fitted above

The confusion matrix accompanying this model is shown in Figure 3-20.

```
     y_train
y_h  0   1
  0  56  17
  1  15  62
```

Figure 3-20. Confusion matrix for classification tree

Regression Trees

The primary goal in this model is to maximize the probability of landing up at a given leaf as a product of the variables being analyzed. We seek to maximize the information we get about the response variable upon each split we approach. This is modeled by

$$I[C:Y]$$

where I is information, C is the variable that determines the leaf we move toward, and Y is the response variable

$$I[Y;A] = \sum_a \Pr(A=a) I[Y;A=a]$$

where,

$$I[Y;A=a] = H[Y] - H[Y|A=a]$$

80

where

$$H(X) = E[I(X)] = E[-\ln(P(X))]$$

Regardless of whether we are looking at continuous or discrete variables, we calculate the sum of squares the same way

$$S = \sum_{x \in leaves(T)} \sum_{i \in c} (y_i - m_c)^2$$

where $m_c = \frac{1}{n_c} \sum y_i, i \in c$, the prediction for leaf c.

Uncertainty in prediction using regression trees, similar to the uncertainty seen in classification trees, is an issue worth considering when employing these models. Primarily, these uncertainties are imprecise estimates of the conditional probabilities. The tree is also actively changing as the response values shift. We would ideally like a measure of how different the tree could have been if we drew a different sample from the same distribution. This can be estimated using *non-parametric bootstrapping*. Assuming data $(x_1, y_1), (x_2, y_2), ..., (x_n, y_n)$, we draw a random set of integers $J_1, J_2, ..., J_n$, independently and uniformly from the numbers 1:n, with replacement. Then we set

$$(X'_i, Y'_i) = (x_{J_i}, y_{J_i})$$

where we treat this bootstrapped sample just like the original data and fit a tree to it. Repeated over many iterations, we get a bootstrap sampling distribution of trees. This approximates the actual sampling distribution of regression trees. The spread of the prediction of our bootstrapped trees around the original indicates the distribution.

Limitations of Decision Trees

Typically, the most difficult parts of building a decision tree are choosing the rule that creates the best decision tree and choosing a tree size that isn't overly complex, which leads to overfitting in the training set, or one that doesn't yield any actionable insights at all. To make things worse, it's difficult to tell when exactly overfitting occurs just from the training error alone. To mitigate these problems, it is generally encouraged that decision trees have a sufficient training example size. Ideally, the model fits to the data reasonably well, and the rules employed to determine splits in direction should not be overly complex. The stopping criterion ultimately controls when we reach a leaf. Examples of often-used rules are to stop when the information yielded decreases below a certain user-determined threshold or when the "child" of the "parent" node yields a sufficiently small enough set of data points. Moving forward from this however, decision trees are relatively simple models that don't always perform very well on complex data with respect to regression problems, and also don't perform well on categorical data where there are multiple levels for each category.

Ensemble Methods and Other Heuristics

For instances in which standard machine learning algorithms fail, a significant boost in accuracy can be achieved from algorithms that are in actuality combinations of multiple algorithms. We refer to these as *ensemble methods*.

Gradient Boosting

Originally developed by Leo Breiman, *gradient boosting* is a technique used on regression and classification problems for the purposes of producing a superior model from weaker models. It builds the model iteratively, and the optimization problem is to minimize the gradient of this function. Let's take a model F, which we expect to predict a value y_h, with the objective of minimizing the squared error. Let M be the number of boosting iterations we want to go under, where $1 \leq m \leq M$. We assume that at the outset of our experiment we will have a model F_m, which we seek to improve. Therefore:

$$F(x)_{m+1} = F(x)_m + h(x) = y$$

$$h(x) = y - F(x)_m$$

Gradient boosting seeks to make F_{m+1} more correct than the previous model. Other loss functions that have been proposed are the squared error loss function given by

$$h(x) = \frac{1}{2n} \Sigma \left(h_{\theta_x}^i - y^i \right)^2$$

where

$$\nabla h(x) = \frac{1}{n} \Sigma \left(h_{\theta_x}^i - y \right)$$

where n = the number of observations within data set X.

Gradient Boosting Algorithm

1. Define the optimization problem as

$$F^* = \arg\min_F E_{x,y}[L(y, F(x))]$$

where $L(y, F(x))$ is some differentiation loss function, such as the gradient of the squared loss as shown earlier.

2. Calculate the residuals as given by the following equation for $m = 1, \ldots, M$:

$$r_{im} = -\left[\frac{\partial L(y_i, F(x_i))}{\partial F(x_i)}\right]_{F(x)=F(x)_{m-1}}$$

3. Use the initial model with a training set to iteratively improve the performance.

4. Calculate γ_m via the following equation:

$$\gamma_m \arg\min_{\gamma} \Sigma_i^n L(y_i, F(x_i)_{m-1} + \gamma h_m(x_i))$$

5. Repeat 2–4 until convergence.

Random Forest

The final ensemble method I will address in this chapter is that of the random forest. Simply stated, *random forests* are combinations of several decision trees, such that each decision tree can be considered unique from the others with respect to the features it evaluates at a given branch. Although the length of these trees is homogenous, each tree's decision is independent from one another. The value we choose for a given observation typically is the average value with respect to all the trees in the case of regression, or it's the average (or most prevalent) observation with respect to all of the trees in classification.

Limitations to Random Forests

Random forests' main limitations is the fact they, similar to the trees they are made of, have a tendency to overfitting. The same techniques I recommended for use on decision trees, such as pruning and preemptively limiting growth, should be used here to limit the probability of a tree overfitting.

Bayesian Learning

Built off of Bayes' theorem, and ultimately employed in many machine learning and natural language processing models, Bayesian learning uses representations of random variables and their conditional dependencies via a directed graph. Bayesian learning is used in situations such as determining the sentiment of a word given the context it is within, and finding the probability of a name being that of a female or a male based on the genders it typically is prescribed to within a test set.

Naïve Bayes Classifier

A simple application of Bayes' theorem is to the case of classification. Naïve Bayes classifier uses conditional probability to determine the likelihood of an event. Let's say we have a vector $z = [z_1, z_1, \ldots, z_3]$ and we want to determine the probability of event A. We would model this equation as

$$P(A|z) = \frac{P(z|A)P(A)}{P(z)}$$

where *P(A|Z)* is defined as the posterior probability, *P(z|A)* is the prior probability, *P(A)* is the likelihood, and *P(z)* is the probability of the instance occurring (this can almost always be ignored). Now we want to use this formula to properly be able to classify observations. To this, we turn this into an optimization problem, given by the following equation:

$$\hat{y} = \arg\max_{k \in \{1, \ldots, K\}} P(A_k) \prod P(z_i | A_k)$$

We assign a value to *y* based on the value that maximizes the probability of some event *A*. Although this isn't the only way to use a naïve Bayes classifier, it's an example of one of the more common ways Bayes' theorem is applied for the purpose of classification.

Limitations Associated with Bayesian Classifiers

Bayesian classifiers' biggest limitation lies mostly the fact that it assumes the independent nature of features, which won't always be the case in many contexts in which we're analyzing data. Once it's established that feature independency doesn't exist, we can't use this classifier at all:

```
#Bayesian Classifier
require(e1071)

#Fitting Model
bayes_classifier <- naiveBayes(y = y_train, x = x_train , data = x_train)
y_h <- predict(bayes_classifier, x_train, type = c("class"))

#Evaluating Model
confusion_matrix <- table(y_h, y_train)
print(confusion_matrix)
```

When executing the preceding code, the confusion matrix shown in Figure 3-21 is yielded.

```
       y_train
y_h  0   1
  0  56  17
  1  15  62
```

Figure 3-21. *Confusion matrix for Bayesian classifier example*

Final Comments on Tuning Machine Learning Algorithms

One of the more difficult parts of practicing machine learning algorithms that I've not addressed yet is the concept of *parameter tuning*. The amount of parameters one can tune depends on which algorithm is being employed, but nonetheless this is a challenge that is noted across the entire discipline. We have discussed why it's important to ensure that overfitting doesn't occur so we achieve as robust a solution as possible. Generally speaking, robustness is reflected by stability of prediction power from one data set to the next, and overfitting is reflected by stark drop-offs in predictive power from on data set to the next. I'll now discuss how to achieve this robustness via methods in the following sections.

50/25/25 Cross-Validation

Users should use a validation set to do the parameter tuning against, which should be 50% of the size of the total data set. Then the users should create two training sets: one will be used to train their tuned algorithm, and the other to test the degree of robustness/check for overfitting. Other percentage splits can be examined to see the difference in performance.

Tune One Parameter at a Time

Should the reader be using the packages and not a custom implementation of an algorithm, there will likely be parameters that are set to default values. Trying to change more than one parameter at a time is difficult not only for the sake of the results of the algorithm being yielded in a timely fashion, but also due to the fact that it's hard to separate the contribution of specific parameters from the degree of change in the output. For example, random forests get a great deal of their power from the largeness of the individual trees as well as from the amount of trees allowed to have within a given model. Augmenting both of these at the same time distorts the ability to which we can properly tune the algorithm as a whole and ultimately can lead to under or overfitting.

Using Search Algorithms to Tune Machine Learning Parameters

Readers who want to take a more advanced approach are advised to pay close attention in Chapter 8 where we discuss in depth search algorithms that can be used to choose machine learning algorithms. Although still a developing area of research, there has been significant success achieved with using GridSearch and other local search algorithms to choose better algorithms by lowering an error statistic of a regression algorithm or increasing an AUC score yielded by a classification algorithm.

Reinforcement Learning

Reinforcement learning differs from supervised learning in that the labels we are fitting against in supervised learning problems are never given. Instead, there is a focus on finding the proper balance between leveraging existing knowledge in the model and knowledge that we want the model to find from the environment which is not already known. Integral to the field of reinforcement learning is this subtopic of *probability theory*. In these type of problems, we assume there is a gambler near a group of slot machines who has to decide which machines to play, how many times to play each machine, and in which order to play the machines. When played, each machine provides a random reward from a probability distribution specific to a given machine. The objective is to maximize the amount of money that the gambler will have taken from this period of gambling. Moving forward, we can generally describe the reinforcement learning problem as one that requires an intelligent exploration of an environment, in reference to the same objectives described in the multi-armed gambler approach.

Distinguishing reinforcement learning from supervised and unsupervised methods is the fact that the actions we take significantly affect the subsequent information we get, hence the emphasis on making the best possible decision upon each iteration of a given algorithm. The basic algorithm is described as following:

- Agent
    1. Execute a given action
    2. Observe a certain outcome
    3. Receive a reward, usually modeled in the form of a scalar
- Environment
    1. Receives action performed by the agent
    2. Outputs an observation as well as a scalar reward

We define *history* as the sequences of observations, rewards, and actions that occur with respect to an agent and a given environment, denoted as $H_t = O_1, R_1, A_1, \ldots, O_{t-1}, R_{t-1}, A_{t-1}$. All subsequent observations, rewards, and actions are influenced by the history as it exists in a given experiment. This is what we define as the *state*, denoted as $S_t = f(H_t)$. The state of the environment is not visible to the agent, so it doesn't allow a bias for the actions an agent may pick. In contrast, the state of the agent is internal. The information also has a state, which is described as a *Markov process*. It should be noted that because this is an introductory book to deep learning, the application of reinforcement learning won't be as in depth as it would be in more advanced books. That said, it is my hope that from reading this book, those who currently find reinforcement learning problems inaccessible will be able to tackle these problems upon attainment of a solid understanding of the concepts addressed during the course of this text.

Summary

We now have reached the end of our review of the necessary components of optimization and machine learning. This chapter, as well as the prior chapter, should be used as a reference point for understanding some of the more complex algorithms we shall discuss in the chapters moving forward. Now, we'll progress into discussing the simplest model within the paradigm of deep learning: single layer perceptrons.

CHAPTER 4

Single and Multilayer Perceptron Models

With enough background now under our belt, it's time to begin our discussion of neural networks. We'll begin by looking at two of the commonest and simplest neural networks, whose use cases revolve around classification and regression.

Single Layer Perceptron (SLP) Model

The simplest of the neural network models, SLP, was designed by researchers McCulloch and Pitts. In the eyes of many machine learning scientists, SLP is viewed as the beginning of artificial intelligence and provided inspiration in developing other neural network models and machine learning models. The SLP architecture is such that a single neuron is connected by many synapses, each of which contains a weight (illustrated in Figure 4-1).

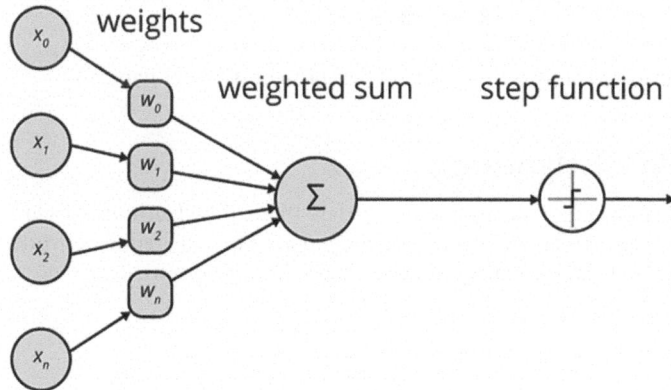

Figure 4-1. *Visualization of single perceptron model*

The weights affect the output of the neuron, which in the example model will be a classification problem. The aggregate values of the weights multiplied by the input are then summed within the neuron and then fed into an activation function, the standard function being the logistic function:

Let the vector of inputs $x = [x_1, x_2, \ldots, x_n]^T$ and the vector of weights $w = [w_1, w_2, \ldots, w_n]$.

The output of the function is given by

$$y = f(x, w^T)$$

where the activation function, when using a logistic function, is the following:

$$f(x) = \frac{1}{1 + e^{-x}}$$

Training the Perceptron Model

We begin the training process by initializing all the weights with values sampled randomly from a normal distribution. We can use a gradient descent method to train the model, with the objective being to minimize the error function. We describe the perceptron model as

$$\hat{y} = f(x, w^T) = \sigma\left(\sum_i^n x_i w_i\right)$$

where

$$\sigma = \frac{1}{1 + e^{-x}},$$

$$\hat{y} = \begin{cases} 1 & \text{if } y \geq \pi^*, \\ 0 & \text{elsewhere} \end{cases}$$

where π^* = the threshold for log odds as described for logistic regression in Chapter 3.

Widrow-Hoff (WH) Algorithm

Developed by Bernard Widrow and Macron Hoff in the late 1950s, this algorithm is used to train SLP models. Though similar to the gradient method used to train neural networks (mentioned earlier), the WH algorithm uses what is called an *instantaneous algorithm*, given by

$$w_i(k+1) = w_i(k) - \eta\left(\frac{\partial E}{\partial w_i}\right) 1$$

$$\frac{\partial E}{\partial w_i} = \frac{1}{2}\sum_{m=1}^{M}1\big(h_{\theta_x}-y(k)\big)\left(-\frac{\partial y(k)}{\partial w_i}\right)$$

$$=\sum_{m=1}^{M}\big(h_{\theta_x}-y(k)\big)\big(-x_i(k)\big)$$

$$=\sum_{m=1}^{M}\delta(k)x_i(k)$$

where

$$\delta(k)=\big(h_{\theta_x}-y(k)\big)$$

We can therefore summarize the preceding equations into the following:

$$w_i(k+1)=w_i(k)+\eta\delta(x)x_i(k)$$

In this manner, we have the same optimization problem that we would in any traditional gradient method. Our goal is to minimize the error of the model by adjusting the weights applied to the inputs of data via gradient descent. With a classification problem in mind, let's use logistic regression as our baseline indicator while also comparing it to a fixed rate perceptron indicator and the bold driver adaptive gradient using the WH algorithm.

Limitations of Single Perceptron Models

The main limitation of the SLP models that led to the development of subsequent neural network models is that perceptron models are only accurate when working with data that is clearly linearly separable. This obviously becomes difficult in situations with much more dense and complex data, and effectively eliminates this technique's usefulness from classification problems that we would encounter in a practical context. An example of this is the XOR problem. Imagine that we have two inputs, x_1 and x_2 for which a response, y, is given, such that the following is true:

x_1	x_2	y
0	0	0
1	0	1
0	1	1
1	1	0

From the following example, we can see that the response variable is equal to 1 when either of the explanatory variables is equal to 1, but is equal to 0 when both explanatory variables are equal to each other. This situation is illustrated by Figure 4-2.

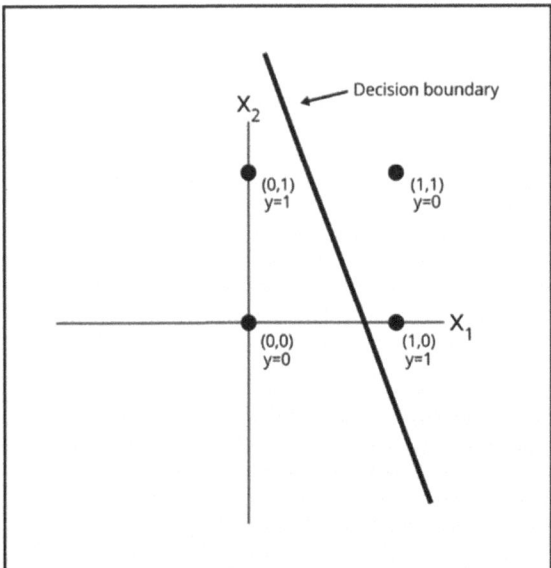

Figure 4-2. *XOR problem*

Let's now take a look at an example using SLP with data that is not rigidly linearly separable to get an understanding of how this model performs. For this example, I've created a simple example function of a single layer perceptron model. For the error function, I used 1 minus the AUC score, as this would give us a numerical quantity such that we could train the weight matrix via back-propagation using gradient descent. Readers may feel free to use the next function as well as change the parameters.

We begin by setting some of the same parameters that we did with respect to our linear regression algorithm performed via gradient descent. (Review Chapter 3 if you need to review the specifics of gradient descent and how it's applied for parameter updating.) The only difference here is that we're using a different error function than the mean squared error used in regression:

```
singleLayerPerceptron <- function(x = x_train, y = y_train, max_iter = 1000, tol = .001){
#Initializing weights and other parameters
  weights <- matrix(rnorm(ncol(x_train)))
  x <- as.matrix(x_train)
  cost <- 0
  iter <- 1
  converged <- FALSE
```

CHAPTER 4 ■ SINGLE AND MULTILAYER PERCEPTRON MODELS

Here, we define a function for a single layer perceptron, setting parameters similar to that of the linear regression via the gradient decent algorithm defined in Chapter 3. As always, we cross-validate (this section of code redacted, please check GitHub) our data upon each iteration to prevent the weights from overfitting. In the following code, we define the algorithm for the SLP described in the preceding section:

```
while(converged == FALSE){
     #Our Log Odds Threshold here is the Average Log Odds
     weighted_sum <- 1/(1 + exp(-(x%*%weights)))
     y_h <- ifelse(weighted_sum <= mean(weighted_sum), 1, 0)
     error <- 1 - roc(as.factor(y_h), y_train)$auc
}
```

Finally, we train our algorithm using gradient descent with the error defined as 1 − AUC. In the following code, we define the processes that we repeat until we converge upon an optimal solution or the maximum number of iterations allowed:

```
#Weight Updates using Gradient Descent
#Error Statistic := 1 - AUC
if (abs(cost - error) > tol | iter < max_iter){
     cost <- error
     iter <- iter + 1
     gradient <- matrix(ncol = ncol(weights), nrow = nrow(weights))
     for(i in 1:nrow(gradient)){
        gradient[i,1] <- (1/length(y_h))*(0.01*error)*(weights[i,1])
      }
(Next section redacted, please check github!)
```

As always, it's useful for readers to evaluate the results of their experiment. Figure 4-3 shows the AUC score summary statistics in addition to the last AUC score with its respective ROC curve plotted:

```
#Performance Statistics
cat("The AUC of the Trained Model is ", roc(as.factor(y_h), y_train)$auc)
cat("\nTotal number of iterations: ", iter)
curve <- roc(as.factor(y_h), y_train)
plot(curve, main = "ROC Curve for Single Layer Perceptron")
}
```

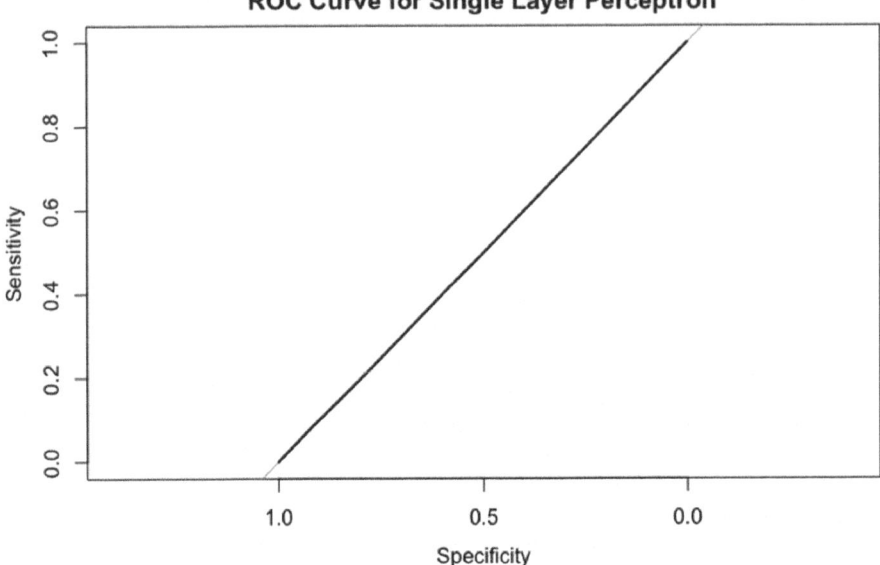

Figure 4-3. *ROC curve*

Summary Statistics

```
      Mean     Std.Dev        Min        Max     Range
1  0.4994949  0.03061466  0.3973214  0.6205357  0.2232143
```

Note that the AUC scores are considerably poor, with the average rating being no better than guessing. Sometimes the algorithm here reaches slightly better results, but this would still likely be insufficient for purposes of deployment. This is likely due to the fact that the classes aren't so clearly linearly separable, leading to misclassification with updates to the weight matrix upon each iteration.

Now that we've seen the limitations of the SLP, let's move on to the successor to this model, the multi-layer perceptron, or MLP.

Multi-Layer Perceptron (MLP) Model

MLPs are distinguished from SLPs by the fact that there are hidden layers that affect the output of the model. This distinguishing factor also happens to be their strength, because it better allows them to handle XOR problems. Each neuron in this model receives an input from a neuron—or from the environment in the case of the input neuron. Each neuron is connected by a synapse, attached to which is a weight, similar to the SLP. Upon introducing one hidden layer, we can have the model represent a Boolean function, and introducing two layers allows the network to represent an arbitrary decision space.

Once we move past the SLP models, one of the more difficult and less obvious questions becomes what the actual architecture of the MLP should be and how this affects model performance. This section discusses some of the concerns the reader should keep in mind.

Converging upon a Global Optimum

By the design of the model, MLP models are not linear, and hence finding an optimal solution isn't nearly as simple as it would be in the case of an OLS regression. In MLP models, the standard algorithm used for training is the back-propagation algorithm, an extension of the earlier described Widrow-Hoff algorithm. It was first conceived in the 1980s by Rumelhart and McClelland and was seen as the first practical method for training MLP networks. It's one of the original methods by which MLP models were trained by using gradient descent. Let E be the error function for the multi-layer network, where

$$E(k) = \frac{1}{2}\sum_{i=1}^{M}\left(h(k)_{\theta_i} - y_i(k)\right)^2$$

We represent the weighted sum value of the individual neurons that is inputted into the hidden layer by the following:

$$s(k)_{h,j} = \sum_{i=1}^{M} w_{h,j,i} x_i(k)$$

Similarly, we represent the output from the hidden layer to the output layer as the following:

$$s(k)_{o,j} = \sum_{i=1}^{H} w_{o,j,i} o_{h,i}(k)$$

With the weights represented by the following:

$$w_{ij}(k+1) = w_{ij}(k) - \eta \frac{\partial E(k)}{\partial w_{ij}}$$

Back-propagation Algorithm for MLP Models:

1. Initialize all weights via sampling from normal distribution.
2. Input data and proceed to pass data through hidden layers to output layers.
3. Calculate the gradient and update weights accordingly.
4. Repeat steps 2 and 3 until algorithm converges upon tolerable loss threshold or maximum iterations have been reached.

After having reviewed this model conceptually, let's look at a toy example. Readers interested in applications of multi-layer perceptrons to practical example problems should pay particular attention to Chapter 10. In the following section of code, we generate new data and display it in the following plot (illustrated in Figure 4-4):

```
#Generating New Data
x <- as.matrix(seq(-10, 10, length = 100))
y <- logistic(x) + rnorm(100, sd = 0.2)

#Plotting Data
plot(x, y)
lines(x, logistic(x), lwd = 10, col = "gray")
```

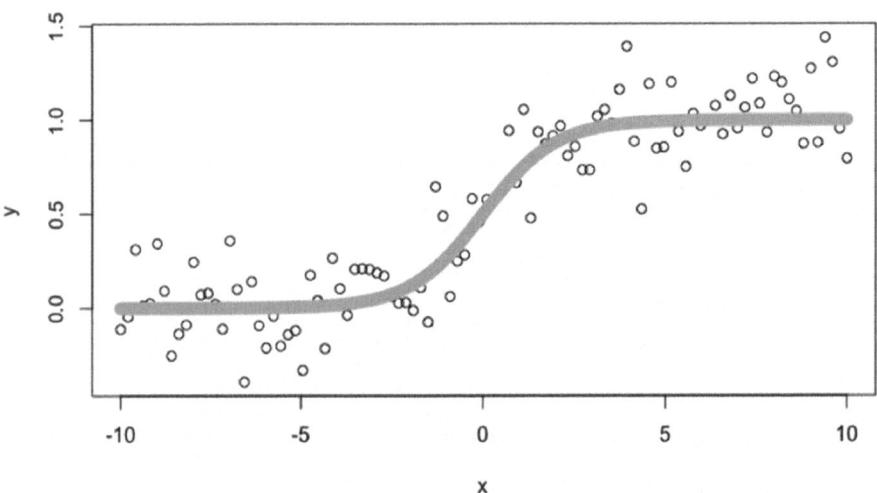

Figure 4-4. Plotting generated data sequence

Essentially, we have a logistic function around which the data is distributed such that there is variance around this logistic function. We then define the variable that holds the weights of the MLP model. I'm using the packaged monmlp, but users may also feel free to experiment with other implementations in packages such as RSNSS and h2o. Chapter 10 covers h2o briefly in the context of accessing deep learning models from the framework:

```
#Loading Required Packages
require(ggplot2)
require(lattice)
require(nnet)
require(pROC)
require(ROCR)
require(monmlp)
```

CHAPTER 4 ■ SINGLE AND MULTILAYER PERCEPTRON MODELS

```
#Fitting Model
mlpModel <- monmlp.fit(x = x, y = y, hidden1 = 3, monotone = 1,
                      n.ensemble = 15, bag = TRUE)
mlpModel <- monmlp.predict(x = x, weights = mlpModel)

#Plotting predicted value over actual values
for(i in 1:15){
  lines(x, attr(mlpModel, "ensemble")[[i]], col = "red")
}
```

When plotting the predictions of the MLP model, we see the results shown in Figure 4-5.

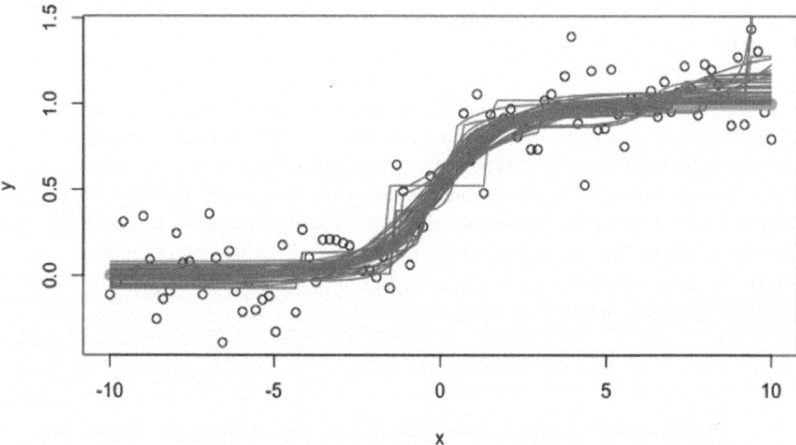

Figure 4-5. *Predicted lines laying over function representing data*

As you can see, there are instances in which the model captures some noise, evidenced by any deviations from the shape of the logistic function. But all the lines produced are overall a good generalization of the logistic function that underlies the pattern of the data. This is an easy display of the MLP model's ability to handle non-linear functions. Although a toy example, this concept holds true in practical examples.

Limitations and Considerations for MLP Models

It is often a problem when using a back-propagation algorithm, where the error is a function of the weights, that convergence upon a global optimum can be difficult to accomplish. As briefly alluded to before, when we are trying to optimize non-linear functions, many local minima obscure the global minimum. We can therefore be tricked into thinking we've found a model which can effectively solve the problem when in fact we've chosen a solution that doesn't effectively reach the global minimum (see Figure 4-6).

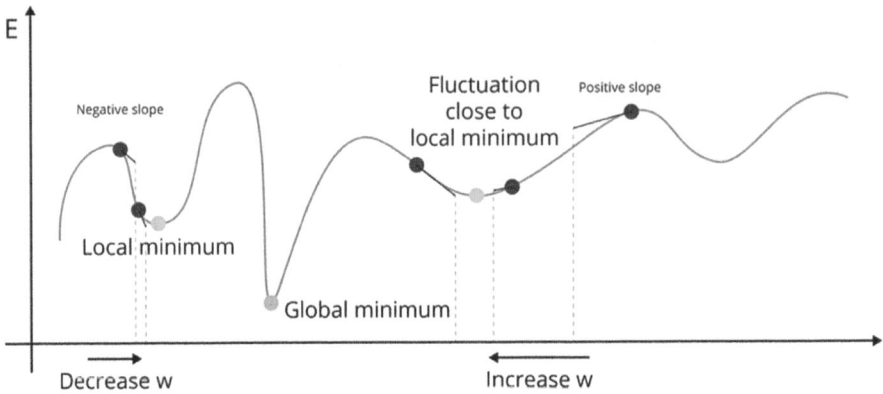

Figure 4-6. Error over weight plot

To alleviate this, the conjugate gradient algorithm is applied. *Conjugate gradient algorithms* differ from the traditional gradient descent method in that the learning rate is adjusted upon each iteration. Many types of conjugate gradient methods have been developed, but all of them have the same motivation underlying them. In the context of the MLP network, we're trying to find the weights that minimize the error function. To do this, we move in the direction of steepest descent, but we change the step size in such a way that it minimizes any possible "missteps" in searching for the global optimum. Let's take a simple example, where we're trying to solve

$$Ax = b$$

where x is an unknown vector, or weights vector in the context of the MLP network, A is the matrix of explanatory variables, and b is the response variable. Now look at the quadratic function

$$f(x) = \frac{1}{2}x^T Ax - b^T + c$$

where c is a constant scalar. When considering an example where A is positive-definite, the optimal solution for minimizing $f(x)$ is the solution to $Ax = b$. When calculating the gradient, we find that $f'(x) = Ax - b$, meaning that the direction of steepest descent would be equal to $b - Ax$. Therefore, we want to adjust the weight vector, x, with the following equation:

$$x_k = x_{k-1} - \eta(b - Ax)$$

The operative part of this method is the transformation of the learning rate, η. By definition, η minimizes the function when the directional derivative of the function with respect to the learning rate is equal to zero. According to the chain rule:

$$\frac{df(x)}{d\eta} = f'(x)^T(-AE), E = y - \hat{y}$$

Finally, we determine the learning rate to therefore be the following:

$$\text{Let } b - Ax = r$$

$$r_k^T r_{k-1} = 0,$$

$$(b - Ax_k)^T r_{k-1} = 0,$$

$$(b - A(x_{k-1} + \eta \, r_{k-1}))^T r_{k-1} = 0,$$

$$(b - Ax_{k-1})^T r_{k-1} - \eta \, (Ar_{k-1}))^T r_{k-1} = 0,$$

$$(b - Ax_{k-1})^T r_{k-1} = \eta \, r_{k-1}^T (Ar_{k-1})$$

$$r_k^T r_{k-1} = \eta \, r_{k-1}^T (Ar_{k-1}),$$

$$\eta = \frac{r_k^T r_{k-1}}{(Ar_{k-1})}$$

How Many Hidden Layers to Use and How Many Neurons Are in It

We typically choose to use hidden layers only in the event that data is not linearly separable. Whenever step, heaveside, or threshold activation functions are utilized, it is generally advisable to use two hidden layers. With respect to using more than one hidden layer, it's largely unnecessary because the increase in performance from using two or more layers is negligible in most situations. In situations where this may not be the case, experimentation by observing the RMSE, or another statistical indicator, over the number of hidden layers should be used as a method of deciding. Often, when adding a layer to a neural network model, this will be simple as editing an argument in a function or, in the case of some deep learning frameworks such as mxnet (featured in later chapters), passing values from a previous layer through an entirely new function. With respect to how many neurons should be within a given hidden layer, this must be tested for with the objective of minimizing the training error. Some suggest that it has to be between the input and output layer size, never more than twice the number of inputs, capturing .70-.90 variance of the initial data set—or to use the following formula:

$$\# \, Hidden \, Units = (\# \, inputs + \# \, outputs) * \frac{2}{3}.$$

CHAPTER 4 ■ SINGLE AND MULTILAYER PERCEPTRON MODELS

Briefly, let's look at the difference between the conjugate gradient training method and traditional gradient descent using the RNSS package in R with the following code:

```
#Conjugate Gradient Trained NN
conGradMLP <- mlp(x = x, y = y,
size = (2/3)*nrow(x)*2,
                 maxit = 200,
                 learnFunc = "SCG")
#Predicted Values
y_h <- predict(conGradMLP, x)
```

We begin by defining the neural network using the `mlp()` function, in which we specifically denote the `learnFunc` argument as SCG (scaled conjugate gradient). We also choose the `size` parameter (the number of neurons in a neural network) using the 2/3 rule mentioned earlier.

Now let's compare the MSE of both the MLP model shown prior and this one we've just constructed:

> MSE for Conjugate Gradient Descent Trained Model: 0.03533956
>
> MSE for Gradient Descent Trained Model: 0.03356279

Although only a slight difference in this instance, we can see that the conjugate gradient method yields a slightly inferior MSE value than the traditional gradient descent method in this instance. As such, it would be wise, given this trend of staying consistent, to pick the gradient descent trained method.

Summary

This chapter serves as an introduction into the world of neural networks. Moving forward, we will discuss models that have been developed for tasks that are generally beyond what SLP and MLP models are made for. Specifically, in Chapter 5, we will look at convolutional neural networks for image recognition as well as recurrent neural networks for time series prediction. Readers who don't feel comfortable yet with the concepts discussed in this chapter are advised to review Chapters 2 though 4 again before advancing to Chapter 5, because many of the concepts referred to in Chapter 5 are addressed at length in those chapters.

CHAPTER 5

Convolutional Neural Networks (CNNs)

Similar to the concepts covered in Chapter 4 with respect to the multi-layer perceptron problem, convolutional neural networks (CNNs) also feature multiple layers used to calculate the output given a data set. This model's development can be traced back to the 1950s, where researchers Hubel and Wiesel modeled the animal visual cortex. At length in a 1968 paper, they discussed their findings, which identified both simple cells and complex cells within the brains of the monkeys and cats they studied. The simple cells, they observed, had a maximized output with regard to straight edges that were observed. In contrast, the receptive field in complex cells was observed to be considerably larger, and their outputs were relatively unaffected by the positions of edges within the aforementioned receptive field. Beyond image recognition, for which CNNs originally gained and still retain their notoriety, CNNs have considerable other applications, such as within the fields of natural language processing and reinforcement learning.

Structure and Properties of CNNs

CNNs are, broadly speaking, multi-layer neural network models. In keeping with the structure of the animal visual cortex as described by Hubel and Weisel, the model can be visually interpreted as shown in Figure 5-1.

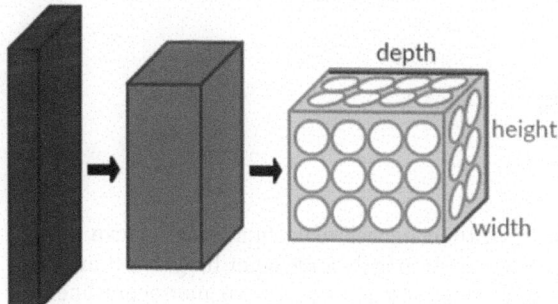

Figure 5-1. *Broad visual display of a CNN*

© Taweh Beysolow II 2017
T. Beysolow II, *Introduction to Deep Learning Using R*, DOI 10.1007/978-1-4842-2734-3_5

Each block represents a different layer of the CNN, which I explain in greater detail later in this chapter. From left to right are the input, hidden (convolutional, pooling, and dropout layers), and fully connected layers. After the final layer, the model outputs a classification. Now consider Figure 5-2.

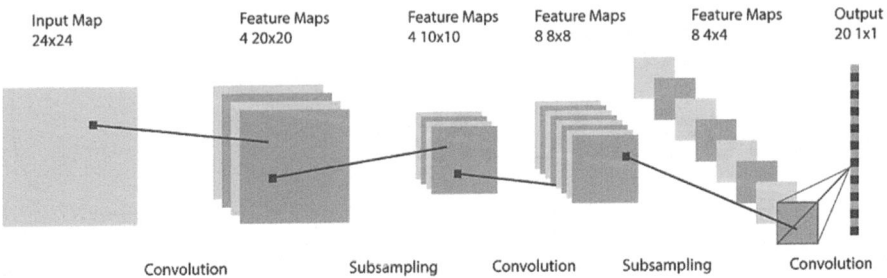

Figure 5-2. *CNN architecture diagram*

Fully connected layers enforce local connectivity between neurons and adjacent layers, as show in Figure 5-2. As such, the inputs of hidden layers are a subset of neurons from the layer preceding that hidden layer. This ensures that the learned subset neurons produce the best possible response. Also, the units share the same weight and bias in the feature/activation map, so that all the neurons in a given layer are analyzing/detecting the exact same feature.

As for *features* in the context of CNNs, I mean portions of an image that are distinct. This is what our filter compares the section of the image it is analyzing to, such that it can determine the degree to which the section of the image being scanned over is similar to the feature being analyzed. Assuming that we have enough training data and enough classes of images, these features are distinct enough that they help to distinguish one class from another.

Imagine we're looking at two images, specifically an X and O, such as in Figure 5-3.

Figure 5-3. *O and X example photo*

If we image both the X and the O as distinct images, to the human eye we can determine them as distinctly different letters. Among their distinguishing factors are that the center of the O is empty, whereas the center of the X features two intersecting lines. Examples of the feature maps of these values when visualized are shown in Figures 5-4 and 5-5.

Figure 5-4. Feature map of "X"

Figure 5-5. Feature map of "O"

These values are often represented as an entry within a matrix with -1 and 1 for black and white respectively. When dealing with color images, each pixel is typically represented as an entry in a matrix with a value of 1 or 256 for black and white respectively. Depending on the language being used, though, zero indexing may affect the representation of RGB values such that the bounds shift backwards by 1.

Components of CNN Architectures

This section covers the components of CNN architectures.

Convolutional Layer

This layer is where the majority of the computation in any given CNN occurs and as such is the first layer after input that an image passes through. Within a convolutional layer, we have filters that scan over a portion of the image. Every filter is not particularly large with respect to height and width, but all of them extend through the entirety of the length of this layer.

For example, imagine we're trying to classify an image as either a 1 or a 0, and the image in actuality is a 1. And imagine that the image has a black background, but the digit is outlined in white pixels. Figure 5-6 shows an example of what this image can be said to look like.

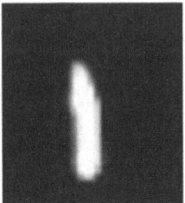

Figure 5-6. Example image of "1"

The computer will distinguish the white pixels as having a value of 1 and the black pixels as having a value of -1. When we input this image through the convolutional layer, the model extracts the unique features of an image, which usually are the colors, shapes, and edges that ultimately define a specific image. Once we have the features of a given training image, we perform what is known as filtering over this inputted image. *Filtering* is the process of taking an image feature, which in this case we can imagine as a 3 x 3 pixel square, and matching it with a patch of that inputted image, which is also a 3 x 3 pixel square. In Figure 5-7, we can see what the process of filtering looks like.

Figure 5-7. Example of a filter

We then multiply the number of the pixel of the feature by the corresponding number of the pixel of the image patch. In the example, we should gain an output of 1 or -1 for each operation. Intuitively, when the pixels match exactly, they should output to 1, and when they don't, they should output to -1. At the end, we take the average of the pixel products. If an image matches exactly, the average should be 1. If it doesn't, it will be considerable degrees lower than one. In this instance, imagine that the image patch and feature selected don't match at all. As such, when we take the average, it should output to -1. We place this product in the center of the position of the image patch we are analyzing with a given feature on what is called a *feature map*. This ultimately will be the output of the convolution layer and will be used in the following layer. The convolutional layer will, over different iterations, produce multiple feature maps. The process of matching a feature with a given image patch over every possible position is known as *convolving* an image. We denote the feature/activation map for a given CNN as

CHAPTER 5 ■ CONVOLUTIONAL NEURAL NETWORKS (CNNS)

$$h_{i,j}^k = \tanh\left(\left(w^k x\right)_{i,j} + b_k\right)$$

where w^k is the weight, b_k is the bias, x is the value of the specific pixel, and tanh is for non-linearities in data. The subscripts i,j refer to the entry of the matrix that represents the feature/activation map. The weight, w^k, is ultimately what connects the pixels in the feature maps to the preceding layer. The convolution layer, ultimately, is a stack of the feature maps that were yielded from the operation described earlier. We then put the feature maps into the pooling layer. We calculate the spatial size of the output volume as

$$Spatial\ Size_{Output} = \frac{W - F + 2P}{S + 1}$$

where W = input volume size, F = size of receptive field of in convolution layer, P = amount of zero padding, and S = stride.

Pooling Layer

Between successive convolutional layers, it's common to place what is called a *pooling layer* in between. Simply stated, the pooling layer takes the feature maps produced in the convolution layer and "pools" them into an image. The pooling layer effectively performs dimensionality reduction, hence the prior emphasis on spatial representation, thereby reducing the complexity of the model. This can be compared to the process of pruning in decision trees and similarly helps to prevent overfitting of a given model. In the prototypical CNN model, the pooling layer has a 2 x 2 filter, a stride of 2, and every depth slice in the input is downsampled such that we move by 2 pixels with respect to height and width. These operations in the pooling layer help to discard 75% of the feature/activation map. This layer uses a max operation, which in the aforementioned example would be taking a max over 4 numbers, or the 4 pixels in any given feature/activation map.

In keeping with the example described earlier, imagine that with a 2 x 2 filter, we're looking at 9 x 9 feature map, where we're analyzing the top lefthand corner with the following scores:

.88	0
0	.95

When using the max operation, we would choose .95, because it's the maximum value within the 2 x 2 window. Because we have a stride of 2, we move 2 pixels to the right, which should mean that we're looking at a 2 x 2 slice of the image where the top lefthand corner of the slice should be the third column of the feature map until we have a max pooled image, which is significantly reduced and therefore removes unnecessary complexities of the model. As a direct consequence of the max operation used in this layer, we needn't be as precise as the prior layer when analyzing the image, and therefore this helps to make a more robust model that can more easily classify inputs. What I mean by this specifically is that the values of the weights connecting each layer can be more generalized to all the training data that they have been exposed to, rather than overfitting in such a manner where the CNN wouldn't perform well out of sample.

The function that determines the spatial size of the output is given by

$$Spatial\ Size_{Output} = W_2 \times H_2 \times L_2,$$

where

$$W_2 = \frac{w_1 - F}{s+1},\ H_2 = \frac{H_1 - F}{S+1},\ L_2 = L_1$$

Rectified Linear Units (ReLU) Layer

Rectifiers are used as another term for an activation function. Typically, we apply the following function to the inputs to this layer

$$f(x) = \max(0,\ x)$$

where x = input to a neuron.

When applied to the feature map, we can imagine that any of the values of the feature map that would be negative now are zero. Specifically, this helps outline the feature map closer to the image it's most associated with. We do this to all of the feature maps to then get a "stack" of images.

Fully Connected (FC) Layer

Any neurons in this layer are connected to all the activation maps in the preceding layer. This layer is usually placed after a user-determined amount of convolutional, pooling, and ReLU layers. The images inputted to this layer will be significantly smaller than the original inputs due to the image reductions specified in the prior operations. In this layer, we scan the reduced images, which should correspond to each feature map, and turn each of the values given here into a list of values. This list then corresponds to one of the k images we put in. Following from the example used in the beginning of the chapter, we originally inputted a 1. After moving through all the layers, we take the average of the scores corresponding to this image, and then this is the probability of the image being a 1 or a 0. It should be noted that the only difference between this layer and the convolutional layer is the fact that the convolutional layers are only connected to a local region in the input and that many of the neurons in a convolutional layer volume share parameters. With this in mind, we can also convert between FC and convolutional layers when constructing a given architecture.

Loss Layer

This layer is where we compare the predicted labels from the actual labels of the images. When trying to classify and object from k possible feature levels, we would use a softmax loss classifier. Using a Euclidean function is also common for the purpose of regressing against the labels of the specific images. Their functions are given by the following:

I. Softmax loss function:

$$\sigma(z)_j = \frac{e^{z_j}}{\sum_{k=1}^{K} e^{z_k}}$$

II. Euclidean loss function:

$$E = \frac{1}{2N} \sum_{i=1}^{N} \|\hat{y}_i - y_i\|_2^2$$

III. Softmax normalization:

$$x'_i = \frac{1}{1 + e^{-\left(\frac{x_i - \mu_i}{\sigma_i}\right)}}$$

When using the back-propagation algorithm, we make a confusion matrix comparing a 1 or a 0, where we subtract the label of the answer to the probability assigned. Following the example that we've been using, let's say 1 = 1, and 0 = 0, but when we input a 1, we only receive a probability of .85 that it is a 0, and a probability of .45 that it is a 1. Therefore, we would have a cumulative error of −.60. We then adjust each feature maps pixel, through the weights displayed in the fully connected layer, using a gradient method as described before, with a designated learning rate. We initialize the weights at 0 and stop the CNN at the point at which the loss tolerance has been reached or the maximum iterations threshold has been reached. The same considerations for convergence upon an optimal solution as described in prior chapters must be taken into consideration.

Tuning Parameters

Images sent to the input layer should be divisible by 2 more than once. Common image dimensions are 32 x 32, 64 x 62, and so on. Convolutional layers should have filters with dimensions of 3 x 3 or 5 x 5 at most, and zero-padding should be performed in such a way that it doesn't alter the spatial dimensions of the input. For the pooling layers, their dimensions should be 2 x 2 with a stride of 2 most often. With these parameters, 75% of the activations will be discarded. Pooling layers that are larger than 3 result in too much loss in the classification process. When describing neurons and their arrangements, hyper-parameters are most relevant to this conversation. Specifically, I will be referring to stride, depth, and zero-padding. Among the most important parameters in CNNs, *stride* is a fixed parameter that determines the number of pixels that slide through a filter. For example, if the stride is 2, then 2 pixels at a time slide through the filters. Typically, stride is no greater than 2, and no less than 1. *Zero-padding* is the size of the zeroes around the border of the input volume. Through controlling zero-padding, we can more carefully control size of the activation maps, and other outputs, from layer to layer. Finally, *depth* refers to the number of filters we choose for a given experiment, each of which is what ultimately searches over each image in the convolutional layer.

Notable CNN Architectures

- *LeNet*: Developed in the 1990s by renowned deep learning researcher Yann LeCun, LeNet is a relatively simple architecture, all things considered. The purpose of this model was originally to classify digits, read zip codes, and perform general simple image classification. This is considered the analogue to a "Hello World" program that any developer first writes in a given language, because it's considered to be the first successful CNN application to a practical task. As Figure 5-8 illustrates, the layers involved are as follows:

    - input, conv layer, ReLU, pooling layer, conv layer, ReLU, pooling layer, fully connected, ReLu, fully connected, and softmax classifier.

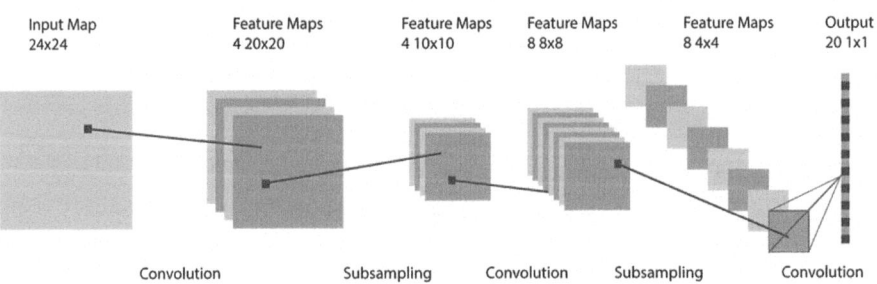

Figure 5-8. Visualization of LeNet

CHAPTER 5 ■ CONVOLUTIONAL NEURAL NETWORKS (CNNS)

- *GoogLeNet (Inception)*: This architecture won the ImageNet Large-Scale Visual Recognition Challenge (ILSVRC) competition in 2014 in homage to Yann LeCun's LeNet. It was developed by Christian Szegedy, Wei Liu, Yangqing Jia, Pierre Sermanet, Scott Reed, Dragomir Anguelov, Dumitru Ehran, Cinven Vanhoucke, and Andrew Rabinovich. The name GoogLeNet is derived from the fact that a considerable number of the developers of the architecture work at Google Inc. In their paper "Going Deeper with Convolutions," they describe an architecture that allows for "increasing the depth and width of the network while keeping the computational budget constraint." As Figure 5-9 illustrates, the structure as proposed is as follows:
    - Input, conv. layer, max pool, conv layer, pooling layer (with max function), inception (2 layers), max pool, inception, inception (5 layers), max pool, inception (2 layers), average pooling layer, dropout, ReLU, softmax classifier.

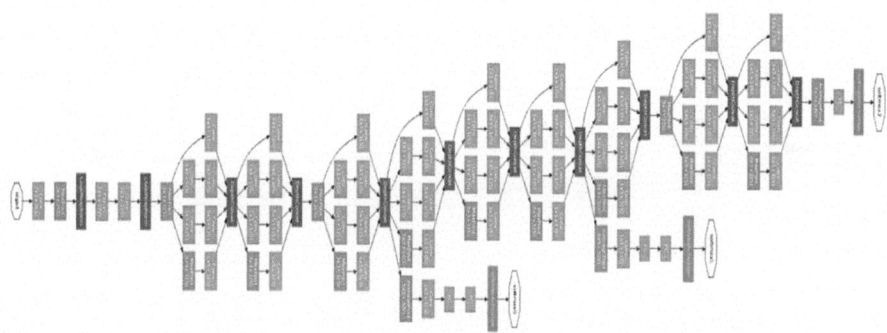

Figure 5-9. *GoogLeNet architecture*

The focus of the inception architecture is that through the orientation in the layers as described earlier, the CNN model allows for "increasing the number of units at each stage" without doing so to the point where the model becomes too complex. Overall, the model seeks to process visual information at various scales and then aggregate the calculations to the next stage so higher levels of abstraction are analyzed simultaneously.

- *AlexNet*: Developed by Alex Krizhevsky, Ilya Sutskever, and Geoffrey Hinton, this won the ILSVRC in 2012. Similar in architecture to LeNet, AlexNet uses "non-saturating neurons" and efficiently implements the GPU for the convolution layers. The neurons in fully connected layers are connected to all neurons in the previous layer, response-normalization layers follow the first and second convolutional layers, and the kernel of layers two, four, and five are connected only to the kernel maps in the previous layer, which would be on the same GPU. The architecture is as follows (and shown in Figure 5-10):

  - Convolutional (5 layers), fully connected (3 layers), [output is 1000-way softmax classifier]

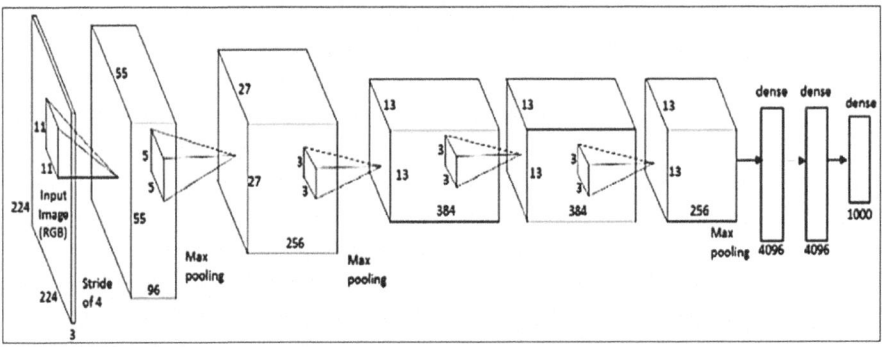

Figure 5-10. AlexNet architecture

- *VGGNet*: This took second place to AlexNet in the ILSVRC 2014 competition. VGGNet was developed by Karen Simonyan and Andrew Zisserman from the University of Oxford. The receptive field is 3 x 3, with 1 x 1 filters, stride is 2, and max pooling size is 2 x 2. The architecture is such that the input is fed through several convolutional layers, to three fully connected layers (the first and second layers have 4096 channels, and the final is a softmax layer that performs 1000-way classification).

- *ResNet*: The first-place winner of ILSVRC 2015, ResNet features 152 layers—far exceeding the amount of the previously mentioned networks. It was developed by Kaimin He, Xiangyu Zhang, Shaoqing Ren, and Jian Sun, all of whom are from Microsoft Research. The purpose of this architecture is to form a network that learns residual functions with references to the layer inputs, rather than a network learning unreferenced functions. The end result is a network that is considerably easier to learn, significantly easier to optimize, and that gains accuracy from increased depth, rather than one that loses accuracy from that depth.

Regularization

When multi-layer perceptrons have more than one layer, they are known to have the ability to approximate a given target, which then would lead to overfitting. To prevent overfitting, *regularizing* the input data is often recommended, however this is a slightly different process in the case of CNNs, we can use: 1) *DropOut*, which is taken from the inspiration of a phenomena observed within the human brain. This is where a given hidden layer has the probability of not being passed through with the probability we set as a hyperparameter. 2) Stochastic pooling, where the activation is picked randomly. *Stochastic Pooling* doesn't require hyper-parameters and can be used as a heuristic, so to speak, with other regularization techniques. 3) *DropConnect*, which is a generalization of dropout, where each connection can be dropped with the probability of 1 - p. Each unit in this layer inputs data from random units in the preceding layer, which change upon every iteration. This helps ensure that the weights don't overfit. 4) Weight Decay, which functions similarly to L1/L2 regularization, where we heavily penalize large weight vectors.

Of these methods, there has been a considerable amount of enthusiasm around using DropOut in CNNs, because it's been shown to be an effective and powerful technique. Beyond preventing overfitting, DropOut has been observed to improve the computational efficiency of networks with large amounts of parameters, as this form of regularization causes a network to in effect become smaller during a given iteration. After all these iterations, the smaller networks' performance can be averaged into a general prediction of what a complete network would have performed as. Secondly, it is observed that the DropOut layer introduces randomized performance in the network that allows noise within the data to be averaged over, such that its masking of signals within the data is diminished.

It's not uncommon to use L1 regularization either, but be aware of the fact that the weight vectors in this instance can often shrink to 0—sometimes enough so that we can be left with a sparsely populated weight matrix. The negative effect of this type of regularization is that the inputs to certain layers that contain important information may become entirely unnoticed due to a "dead" connection between layers. In contrast, though, when you feel specifically that you want very explicit feature selection, L1 regularization may yield significantly better performance.

L2 regularization is traditionally seen as the standard method by which regularization is performed in CNNs, because it tends to penalize abnormally large weights and favor those that are generally mild in their proportion relative to the entirety of the matrix. In contrast to L1 regularization, you get a considerably more populated weight matrix, which will cause the network to feed more data from a given layer to the next. As such, feature selection will be less stark than when using L1 regularization.

The final type of regularization you should know about would be an addendum to either L1 or L2 regularization via enforcing limits on a given weight's norm size. As such, this would allow the parameter updates to have a hard limit and therefore limit the number of possible solutions a given network can yield. This would help to train the network faster via the limitation of possible solutions, and in the optimal solution prevent the parameters from updating too far in the incorrect direction.

Summary

We have sufficiently covered the concepts of CNNs and walked through all the architectures that are most recent at this time. See Chapter 11 for an applied example of a CNN, specifically with respect to the preprocessing of image data—a highly important step in the constructing of image recognition software. Moving forward, we will discuss recurrent neural networks (RNs) and the intricacies of detecting patterns in time series-based data.

CHAPTER 6

Recurrent Neural Networks (RNNs)

Recurrent neural networks (RNNs) are models that were created to tackle problems within the scope of pattern recognition and are fundamentally built on the same concepts with respect to feed-forward MLPs. The difference is that although MLPs by definition have multiple layers, RNNs do not and instead have a directed cycle through which the inputs are transformed into outputs. I'll begin the chapter by covering several RNN models and end it with a practical application of RNNs.

Fully Recurrent Networks

Imagine that we have an input, x, that we're inputting into an RNN model, where we define the state as h, with the inputs being multiplied by a weight matrix, W. So far, everything is the same as it would be in previously described neural network models—but as stated before, RNNs perform the same task on the inputs over time. Because of this, to calculate the current state of a neural network, we derive the following equation:

$$h_t = f(W_t x + W_R h_{t-1} + b), \text{ where } f = tanh, ReLU$$

$$y_t = f(W h_t + b), \text{ where } W_{t,r} = weights, \ h_t = hidden\ layer, \ b = bias$$

The key characteristic here is that when the neural network performs these operations, it "unfolds" into multiple new states, each of which is dependent on the prior states. Because these networks perform the same task for every input that's put in, in addition to the functional dependency of the model, RNNs are often referred to as *having memory*. Figure 6-1 illustrates an RNN.

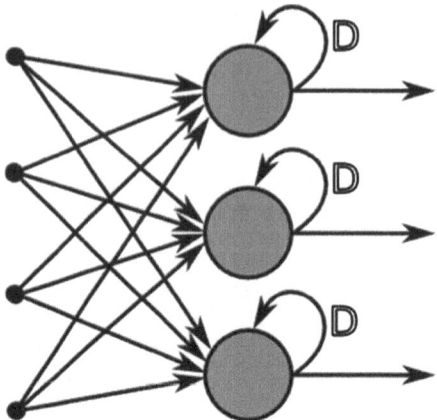

Figure 6-1. *Architecture of recurrent neural network*

This form of the RNN was developed in the 1980s. Similar to other neural networks, multiple layers of neurons are connected by weights, with each weight being altered via back-propagation methods. We alter our weights based on an evaluation statistic, which in this case is the weighted sum of the activation units at a given time step. The total error is the sum of all these individual weighted sums across all time steps. There may be teacher-driven target activations for some of the output units at certain time steps. For example, if the input sequence is a speech signal corresponding to a spoken digit, the final target output at the end of the sequence may be a label classifying the digit. For each sequence, its error is the sum of the deviations of all target signals from the corresponding activations computed by the network. For a training set of numerous sequences, the total error is the sum of the error of all individual sequences.

Training RNNs with Back-Propagation Through Time (BPTT)

Sepp Hochreiter and Jurgen Schmidhuber, among others, are considered among the greatest pioneers for development of training methods for deep learning. The standard method is called *back-propagation through time* (BPTT). BPTT is roughly the same as regular back-propagation, except it was created to deal with a specific problem that RNNs have, which is the fact that we are evolving a model through various time steps. For each training epoch, we begin by first training on reasonably small sequences and gradually increasing the length of the aforementioned training sequence. Intuitively, this is typically envisioned as training on a sequence of length 1,2, through N, where N is the maximum possible length of the sequence. Here is an equation describing this phenomenon more succinctly

$$\delta_{p,j}(t-1) = \sum_{h}^{m} \delta_{ph}(t) u_{hj} f'(s_{pj}(t-1))$$

where t = time step, h = index for hidden node at t, j = hidden node at step t = 1, and δ = errors.

In detail, we can view the phenomena as the following: we define W as the matrix of weights for the output layer with the equation

$$W(t+1) = W(t) + \eta s(t) e_o(t)^T$$

where e_o = errors from the output layer:

$$e_o(t) = d(t) - y(t)$$

We now have k sequences, through which we unfold the network into a regular feed-forward network that we've been observing up until this point. However, the recurrent layer in RNN model simultaneously takes the input from the preceding layer as well as the successive layer. To offset the change in weights that occurs from simultaneous inputting when back-propagating the errors, we average the updates that each layer receives.

Elman Neural Networks

RNN architectures received additional contributions from Jeffrey Elman, who is credited with creating the Elman network model named after him. Primarily, the architectures Elman constructed were for language-processing algorithms, but they can also be useful for any problem in which the input data is sequential or time-series based. Figure 6-2 shows the basic structure of an Elman network.

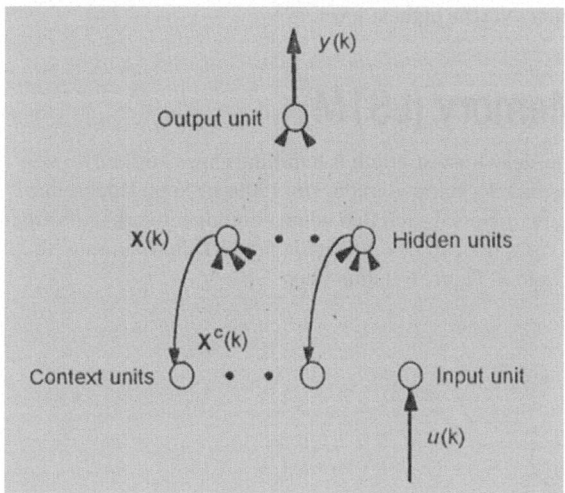

Figure 6-2. *Illustration of Elman network*

Elman included a layer of context units in this architecture that are distinguished by the fact that their functionality is highly concerned regarding prior internal states. One of the key distinctions of an Elman network is for the output of the hidden layer to feed into the context units in the preceding layer as well, but the weights that connect the context units and hidden layer have a constant value of 1, making the relationship linear. After this, the input layer and context layer simultaneously activate the hidden layer, whereupon the hidden layer also outputs a value while performing the update step. During the next epoch, the training sequence previously described occurs the same way, except here we observe that the layer with the context units now adopt the values of the hidden layer from the prior epoch. This feature of the context units colloquially has been described as the network *having memory*. Training this neural net requires a multitude of steps, the number of which ultimately depends on the length of the string being chosen.

Neural History Compressor

The *vanishing gradient* specifically refers to the gradients in earlier layers of a network becoming infinitesimally small. This occurs due to whatever activation function we use, usually a tanh or sigmoid. Because these activation functions "squash" the inputs into relatively small ranges to make interpolating the results easier, it makes deriving the gradients significantly more difficult. Repeat this process of squashing inputs after multiple stacked layers, and by the time we back-propagate to the first layer, our gradient has "vanished." The problem of vanishing gradients was partially dealt with via the creation of *neural history compressors*—an early generative model implemented as an unsupervised "stack" of recurrent neural networks. The input level learns to predict its next input from the previous input history. In the next higher-level RNN, the inputs are comprised of only the unpredictable inputs of a subset of the RNNs in the stack, which ensures that the internal state is recomputed rarely. Each high-level RNN thus learns a compressed representation of the information in the RNN below. By design, we can precisely reconstruct the input sequence from the sequence representation at the highest level. When we're using sequential data with considerable predictability, supervised learning can be utilized to classify substantially deeper sequences via the highest-level RNN.

Long Short-Term Memory (LSTM)

LSTM is an increasingly popular model whose strength is handling gaps of unknown size between signals in the noise of the data. Developed in the late 1990s by Sepp Hochreiter and Jurgen Schmidhuber, LSTMs are universal such that when enough network units are present, anything a computer can compute can be replicated with LSTMs, assuming we have a properly calibrated weight matrix. Figure 6-3 illustrates.

CHAPTER 6 ■ RECURRENT NEURAL NETWORKS (RNNS)

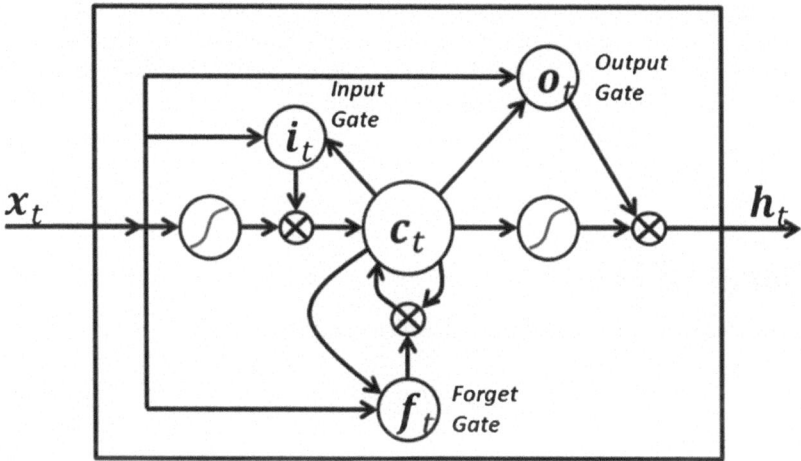

Figure 6-3. *Visualization of long short-term memory network*

The range of applications of LSTMs explains their popularity in part, as they are often used in the fields of robot control, time series prediction, speech recognition, and other tasks. In contrast to the units that we often see in other RNN architectures, LSTM networks contain blocks. Other key distinguishing factor of LSTMs is being able to "remember" a given value for extended periods of time and the gates within the model determining several attributes of the input sequence. Among the considerations of the gates are input significance, when should memory be kept or "garbage collection" occur and data be removed, and output value time. A typical implementation of an LSTM block is shown in Figure 6-3. The sigmoid units in a standard LSTM contain the equation

$$y = s\left(\sum_{i=1}^{N} w_i x_i\right),$$

where s is a squashing function (in many cases, often a logistic function or any activation function, as described in prior models). Looking at Figure 6-3, the sigmoid unit furthest to the left feeds the input to the LSTM block's "memory." From this point forward, the other units in the figure serve as the gates, which either permit or deny access into the LSTM memory. The unit entitled i, which we denote as the input gate of the diagram, will block all values from entering the memory that are very small (close to zero). The *forget gate*, the unit at the bottom of the figure, "forgets" whatever value it was remembering and discards this from the memory. The unit in the top righthand corner of the figure is the "output gate," which determines whether the value stored in the memory of the LSTM should be outputted. Occasionally, we observe units that are denoted by the following symbols: Π or Σ. Units that have the summation symbol are fed back into the LSTM block as to facilitate remembrance of the same value over many time steps sans value decay. Typically this value is also inputted into the three gate units improve their respect decision making processes. The Haramard product, or entrywise product of matrices used in LSTMs, is given by the following in index notation:

$$(A \circ B)_{i,j} = A_{i,j} \times B_{i,j}$$

Traditional LSTM

Above, we have the layers of an LSTM through which our data passes

$$f_t = \sigma_g\left(W_f x_t + U_f h_{t-1} + b_f\right),$$

$$i_t = \sigma_g\left(W_i x_t + U_i h_{t-1} + b_i\right),$$

$$o_t = \sigma_g\left(W_o x_t + U_o h_{t-1} + b_o\right),$$

$$c_t = f_t \circ c_{t-1} + i_t \circ \sigma_c\left(W_c x_t + U_c h_{t-1} + b_c\right),$$

$$h_t = o_t \circ \sigma_h\left(c_t\right),$$

where x = input vector, h_t = output vector, c_t = cell state, (W, U, b) = paramter matrices and vector, $(f_t, i_t,$ and $o_t)$ = remembered information, acquired information, and output, respectively, σ_g = sigmoid function, σ_c = original hyperbolic tangent, σ_h = original hyperbolic tangent.

Training LSTMs

BPPT is used for LSTMs, but due to special features of the LSTM, we can also use gradient descent via BP as we would traditionally. Vanishing gradients in LSTMs are handled specifically by the *error carousel*. LSTMs "remember" their back-propagated errors, which are then fed back to each of the weight. Thus, regular back-propagation is effective at training an LSTM block to remember values for very long durations of time.

Structural Damping Within RNNs

If we're using a conjugate gradient method and it strays too far from the original x, the curvature estimate becomes inaccurate and we may observe an inability to converge upon the global optimum. Suggested by Martens and Sutskever, *structural dampening* is recommended when using conjugate gradient methods. With this method, we penalize large deviations from x, where the formula is given by

$$\tilde{f}_d(x+\Delta x) = \tilde{f}(x+\Delta x) + \lambda \|\Delta x\|^2,$$

where $\|\Delta x\|^2$ is the magnitude of the deviation. λ, similar to ridge regression, serves as a tuning parameter.

The tuning parameter is adaptive and is chosen via a process similar to that of the Levenburg-Marq algorithm described in Chapter 3. It is suggested that we find a reduction ratio, given by the following equation:

$$\rho \equiv \frac{f(x+\Delta x) - f(x)}{\tilde{f}(x+\Delta x) - \tilde{f}(x)}$$

Tuning Parameter Update Algorithm

Weights are updated at each time step and as such augmenting the value in this matrix can cause drastic changes in the output:

$$If\left(\rho > \frac{3}{4}\right)\{$$

$$\lambda \to \frac{2}{3}\lambda.$$

$$\} \, Else \, If\left(\rho < \frac{1}{4}\right)\{$$

$$\lambda \to \frac{3}{2}\lambda$$

$$\} \, Else \, If\left(\frac{1}{4} < \rho < \frac{3}{4}\right)\{$$

$$\lambda \to \lambda, (no \, update)$$

CHAPTER 6 ■ RECURRENT NEURAL NETWORKS (RNNS)

Practical Example of RNN: Pattern Detection

Let's take the example of trying to predict time series based sequential data. In this instance, we're going to try and predict the production of milk at different times of the year (Figures 6-4 and 6-5). Let's begin by examining our data to get an understanding of it:

```
#Clear the workspace
rm(list = ls())
#Load the necessary packages
require(rnn)

#Function to be used later
#Creating Training and Test Data Set
dataset <- function(data){
  x <- y <- c()
  for (i in 1:(nrow(data)-2)){
    x <- append(x, data[i, 2])
    y <- append(y, data[i+1, 2])
  }
  #Creating New DataFrame
  output <- cbind(x,y)
  return(output[1:nrow(output)-1,])
}
```

When working with time series data, we will have to perform a significant amount of data transformation. Particularly, we must create X and Y variables that are slightly different from the given data. From the dataset() function, we create a new X variable, which is time step t, from the original Y variable. We make a new Y variable that is t + 1 from the original Y variable. We then truncate the data by one row such that we remove the missing observation. Moving forward, let us load and visualize the data (shown in Figures 6-4 and 6-5):

```
#Monthly Milk Production: Pounds Per Cow
data <- read.table("/Users/tawehbeysolow/Downloads/monthly-milk-production-pounds-p.csv", header = TRUE, sep = ",")
#Plotting Sequence
plot(data[,2], main = "Monthly Milk Production in Pounds", xlab = "Month", ylab = "Pounds",
     lwd = 1.5, col = "cadetblue", type = "l")
#Ploting Histogram
hist(data[,2], main = "Histogram of Monthly Milk Production in Pounds", xlab = "Pounds", col = "red")
```

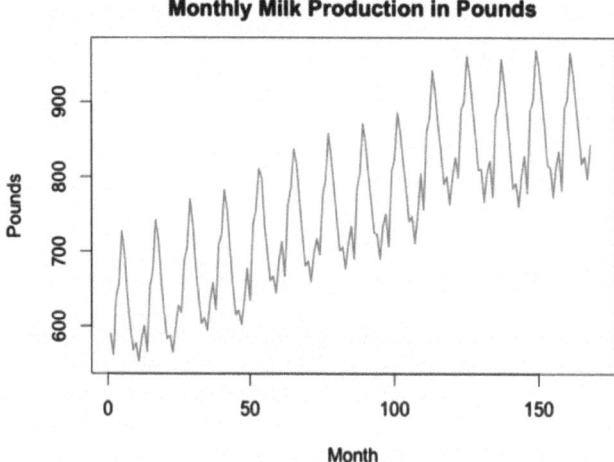

Figure 6-4. Visualization of sequence

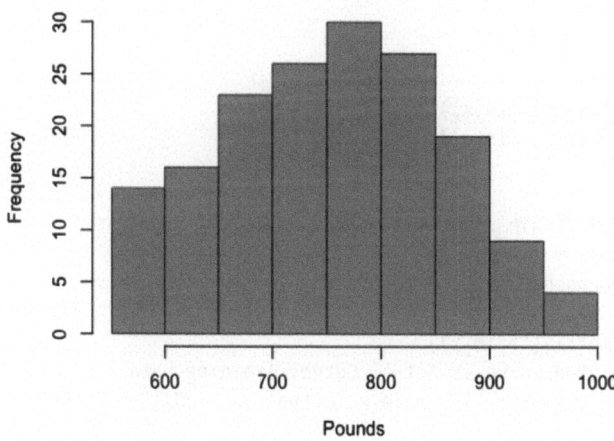

Figure 6-5. Visualization of milk data via histogram

As you can see, our data has a heavy right skew with respect to the frequency of values, despite the seemingly wide range of values.

Now that you've visually understood our data, let's move on and prepare our data to be inputted into the RNN:

```
#Creating Test and Training Sets
newData <- dataset(data = data)

#Creating Test and Train Data
rows <- sample(1:120, 120)
trainingData <- scale(newData[rows, ])
testData <- scale(newData[-rows, ])
```

I recommend that all users use max-min scaling prior to inputting their data into an RNN, because it significantly helps with reducing the errors from a given neural network. Similar to standard normalization, max-min scaling significantly reduces the range of your input data set, but it does so by classifying observations between 0 through 1 rather than by returning how many standard deviations away from the mean the data is. After we have performed this step, we can input our data. Users may feel free to experiment with the parameters, but I have trained the network for good performance.

Now let's evaluate our training and test results (shown in Figures 6-6 and 6-7):

```
#Max-Min Scaling
x <- trainingData[,1]
y <- trainingData[,2]

train_x <- (x - min(x))/(max(x)-min(x))
train_y <- (y - min(y))/(max(y)-min(y))

#RNN Model
RNN <- trainr(Y = as.matrix(train_x),X = as.matrix(train_y),
learningrate = 0.04, momentum = 0.1,
network_type = "rnn", numepochs = 700, hidden_dim = 3)

y_h <- predictr(RNN, as.matrix(train_x))
#Comparing Plots of Predicted Curve vs Actual Curve: Training Data
plot(train_y, col = "blue", type = "l", main = "Actual vs Predicted Curve",
lwd = 2)
lines(y_h, type = "l", col = "red", lwd = 1)
cat("Train MSE: ", mse(y_h, train_y))


#Test Data
testData <- scale(newData[-rows, ])
x <- testData[,1]
y <- testData[,2]
test_x <- (x - min(x))/(max(x)-min(x))
test_y <- (y - min(y))/(max(y)-min(y))
y_h2 <- predictr(RNN, as.matrix(x))
```

CHAPTER 6 ■ RECURRENT NEURAL NETWORKS (RNNS)

```
#Comparing Plots of Predicted Curve vs Actual Curve: Test Data
plot(test_y, col = "blue", type = "l", main = "Actual vs Predicted Curve",
lwd = 2)
lines(y_h3, type = "l", col = "red", lwd = 1)
cat("Test MSE: ", mse(y_h2, test_y))
```

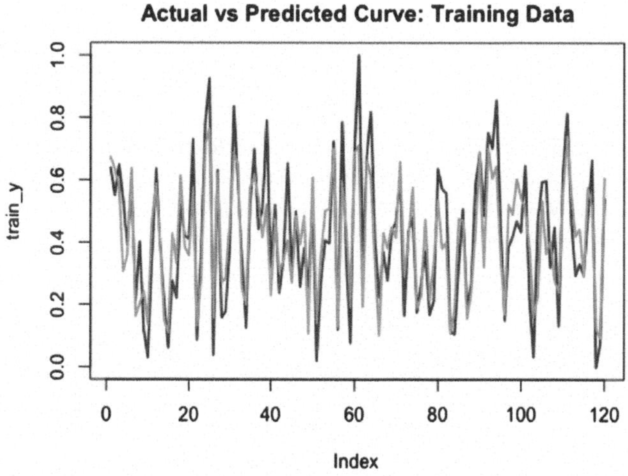

Actual vs Predicted Curve: Training Data

Figure 6-6. *Training data performance*

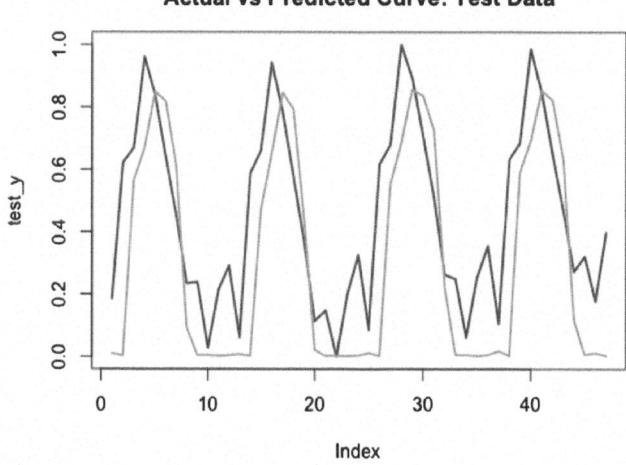

Figure 6-7. *Test set performance*

Respectively, the training and test set have MSEs of 0.01268307 and 0.06666131. Although the MSE for the training set is lower, this is likely just because the training set is significantly larger than the test set. We can see how the test performance is less accurate than the training set by visually comparing the curves in the respective plots. As you can see, the actual curve in both the training and test sets exhibits higher variance than the RNN can completely capture. If you're reading the e-book, the actual curve is in blue and the predicted curve is in red.

Summary

This chapter has effectively covered the most frequently mentioned RNN examples. It also walked the reader through tackling time series data problems. Chapter 7 addresses some of the most recent developments in deep learning and also explores how we can use these insights to tackle even more difficult problems.

CHAPTER 7

Autoencoders, Restricted Boltzmann Machines, and Deep Belief Networks

This chapter covers some of the newer and more advanced deep learning models that have been gaining popularity in the field. It is intended to help you understand some of the recent developments in the field of data science. To see how these models are applied in a practical context, see Chapters 10 and 11, where we will be utilizing these in practical examples.

Autoencoders

Prior to discussing restricted Boltzmann machines (RBMs), I want to address a set of related algorithms. *Autoencoders* are known as feature extractors, in that they are able to learn the encoding/representation of data. The data inputted to an RBM would be the same data that we would input to any machine learning algorithm, but for the sake of simplicity we can imagine it as an M x N matrix where each column is a unique feature and each row a unique observation of N features. It is an unsupervised learning method that uses back-propagation to find a way to reconstruct its own inputs. Developed by Geoffrey Hinton, along with other researchers, autoencoders address the problem of how to perform back-propagation without explicitly telling the autoencoder what to learn from.

Autoencoders consist of two parts: the encoder and the decoder. Let's look at a simple example of what we will denote as an *n/p/n autoencoder architecture*. This architecture is denoted by $n, p, m, \mathbb{G}, \mathbb{F}, \mathcal{A}, \mathcal{B}, \mathcal{X}, \Delta$, where the following are true:

1. $\mathbb{G}$ and $\mathbb{F}$ are sets.

2. n and p are positive integers where $0 < p < n$.

3. Let $\mathcal{A}$ be a function where $\mathcal{A}: \mathbb{G}^p \to \mathbb{F}^n$.

4. Let $\mathcal{B}$ be a function where $\mathcal{B}: \mathbb{F}^n \to \mathbb{G}^p$.

CHAPTER 7 AUTOENCODERS, RESTRICTED BOLTZMANN MACHINES, AND DEEP BELIEF NETWORKS

5. $\mathcal{X} = \{x_1, \ldots, x_M\} \in \mathbb{F}^n$ and when targets are present,
    $\mathcal{Y} = \{y_1, \ldots, y_M\} \in \mathbb{F}^n$.

6. Δ is an L_p norm or some other loss/dissimilarity function.

For any $A \in \mathcal{A}$, and $B \in \mathcal{B}$, the autoencoder transforms the input x into an output vector:

$$\hat{x} = A \circ B(x) \in \mathbb{F}^n$$

Broadly, the problem we seek to solve by using an autoencoder is ultimately an optimization problem—in this case, it is to minimize the loss/dissimilarity function. We define this problem as the following:

$$\min E(A,B) = \min_{A,B} \sum_{m=1}^{M} E(x_m) = \min_{A,B} \sum_{m=1}^{M} \Delta(A \circ B(x_m), x_m)$$

When targets are present:

$$\min E(A,B) = \min_{A,B} \sum_{m=1}^{M} E(x_m, y_m) = \min_{A,B} \sum_{m=1}^{M} \Delta(A \circ B(x_m), y_m)$$

Linear Autoencoders vs. Principal Components Analysis (PCA)

For this example, let's look at the similarities between principal components analysis (PCA) and linear autoencoders. The primary focus of PCA is to find the linear transformations of the original data set that contain the most variability within them in. When translating this analysis to the original data set, we use this to achieve dimensionality reduction. Chapter 8 talks about PCA in greater detail, but I will explain the relation it has to linear autoencoders here. Plainly stated, *PCA is an orthogonal linear transformation where we seek to maximize the variance within each principal component subject to the constraint that each principal component is uncorrelated with each other.* Let us define y as the following:

$$y_i = Ax_i,$$

where $x \in \mathbb{R}^n$ and is the data set, and $A \in \mathbb{R}^{n \times n}$ and is the orthogonal covariance matrix. As is the case with PCA, each principal component should be listed in order of decreasing variance. We define the direction of maximum variance as the following:

$$\hat{w} = \arg\max_{w} \frac{w^T X^T X w}{w^T w}$$

CHAPTER 7 ■ AUTOENCODERS, RESTRICTED BOLTZMANN MACHINES, AND DEEP BELIEF NETWORKS

This by definition is a constrained optimization problem, solvable by using Lagrangian multipliers. Therefore, we can remodel the problem as

$$\mathcal{L}(w, \lambda) = w^T C w - \lambda(w^t w - 1),$$

$$Cw - \lambda w = 0,$$

$$Cw = \lambda w$$

where $C = X^T X$.

Single layer autoencoders will yield almost the exact same eigenvectors as PCA. That said, PCA assumes a linear system in its derivation in contrast to autoencoders that don't. In the instance that we force linearity in an autoencoder, a similar answer will be reached.

To see applications of autoencoders, see Chapter 11, where we specifically use these models for anomaly detection and improving model performance for standard machine learning models.

Restricted Boltzmann Machines

In the 1980s, Geoffrey Hinton, David Ackley, and Terrence Sejnowski developed this algorithm, which can be described as a type of stochastic neural network. At the time, it represented a breakthrough in the science of deep learning because it was among the first models to be able to learn the internal representations of data and have an ability to solve difficult combinatorics problems. The standard restricted Boltzmann machine has a binary-valued hidden and visible unit, consisting of a matrix of weights, W, associated with the connection between a given set of hidden units and visible units, and a bias weight. The hidden, visible, and bias units can be thought as analogous to those same units that appear in a multilayer perceptron model. Given these, the energy of a configuration is stated as the following:

$$E(v,h) = -\sum_{i=1}^{N} a_i v_i - \sum_{j=1}^{N} b_j h_j - \sum_{i=1}^{N}\sum_{j=1}^{N} v_i w_{i,j} h_j$$

This energy function is similar to the output neurons of a Hopfield network (see Figure 7-1), which is a particular type of RNN. Created in the 1980s by John Hopfield, the inputs, as with other RNN models, typically would be data that we suspect to have some underlying pattern (a time series for example). The weighted sum of all inputs is calculated, whereupon it is inputted into a linear classifier such as a logistic function. We define the output as the following:

$$\hat{y} = \begin{cases} 1, & \sum w_i x_i \geq 0 \\ -1, & \sum w_i x_i < 0 \end{cases}$$

CHAPTER 7 ■ AUTOENCODERS, RESTRICTED BOLTZMANN MACHINES, AND DEEP BELIEF NETWORKS

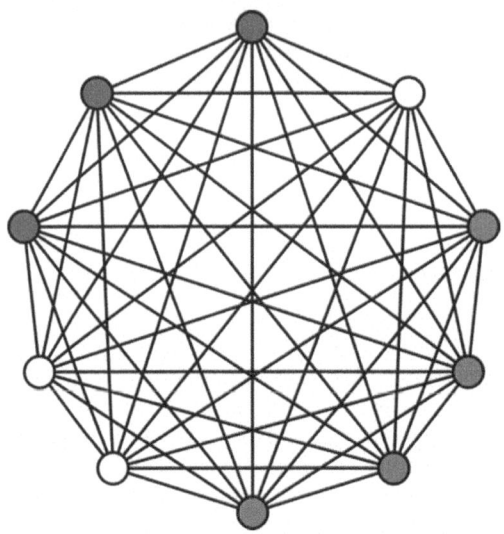

Figure 7-1. Visualization of a Hopfield network

After data is inputted to the model, all the nodes in the network receive specific values. The network is then subjected to a number of iterations using asynchronous or synchronous updating. After a stopping criterion is reached, the values within the neurons are displayed. The primary motivation for Hopfield networks is to discover the patterns stored in the weight matrix.

When referring back to the RBM model, the probability distributions that underlie the data are defined as

$$P(v,h) = \frac{1}{Z} e^{-E(v,h)}, \; Z = \sum e^{-E(v,h)},$$

$$P(v) = \frac{1}{Z} \sum e^{-E(v,h)}$$

where $e^{-E(v,h)}$ = the exponential function, and the superscript is the negative value of the energy function previously described.

RBMs and bipartite graphs share similar properties. As such, the activations from the hidden units are mutually independent given the activations from the visible units such that

$$P(v|h) = \prod_{i=1}^{N} P(h_j|v), \; P(h|v) = \prod_{j=1}^{N} P(h_j|v),$$

CHAPTER 7 ■ AUTOENCODERS, RESTRICTED BOLTZMANN MACHINES, AND DEEP BELIEF NETWORKS

and the individual activation probabilities are

$$P(h_j = 1|v) = \sigma\left(b_i + \sum_{j=1}^{N} w_{i,j} v_i\right), \ P(v_i = 1|h) = \sigma\left(a_j + \sum_{i=1}^{M} w_{i,j} h_i\right),$$

$$\sigma = \frac{1}{1 + e^{-k(x-x_0)}}$$

where a = activation unit.

The values of the visible units of an RBM can be derived from a multinomial distribution, whereas the values of the hidden units are derived from a Bernoulli distribution. In the instance that we use a softmax function for the visible units, we have the following function:

$$P(v_i^k = 1|h) = \frac{exp\left(a_i^k + \sum_{j=1}^{} W_{i,j}^k h_j\right)}{\sum_{k=1}^{K} exp\left(a_i^k + \sum_{j=1}^{} W_{i,j}^k h_j\right)}$$

The optimization of the weights inside an RBM is performed traditionally by using gradient descent via back-propagation until we've converged upon an optimal solution. One of the most popular use cases for RBMs has been to populate missing values within a data set, specifically in the case of collaborative filtering. Chapter 11 looks at a simple example of performing collaborative filtering. If you're interested in reading about performing this with RBMs, search for the paper by Salakhutdinov et al. on using RBMs for collaborative filtering (http://www.machinelearning.org/proceedings/icml2007/papers/407.pdf).

With respect to implementations of RBMs, there are a few packages that you may feel free to explore, such as deepnet, darch, and other implementations online. If you feel advanced enough, you may also seek to create your own implementation. In the meantime, you should check for updates to deep learning frameworks to see if/when they add RBM implementations.

Contrastive Divergence (CD) Learning

Developed by Hinton, *contrasting divergence* (CD) learning is a standard method of training restricted Boltzmann machines. It's based on the idea of using a Gibbs sampling, run for k steps, where it is initialized with a training example of the training set and yields the sample after k steps. It has broader applications as a training method for undirected

CHAPTER 7 ■ AUTOENCODERS, RESTRICTED BOLTZMANN MACHINES, AND DEEP BELIEF NETWORKS

graph models, but its most popular use case is the training of RBMs. I'll begin this discussion by defining the gradient of the log likelihood:

$$\sum_h p(h|v)\frac{\partial E(v,h)}{\partial w_{i,j}} = \sum_h p(h|v)h_i v_j = \sum_h \prod_{k=1}^n p(h_k|v)h_i v_j$$

$$= \sum_{h_i}\sum_{h_{-i}} p(h_i|v)p(h_i|v)h_i v_j$$

$$= \sum_{h_i} p(h_i|v)h_i v_j \sum_{h_{-i}} p(h_{-i}|v) = p(H_i = 1|v)v_j$$

$$= sig\left(\sum_{j=1}^m w_{i,j}v_j + c_i\right)$$

Intuitively, we define the log-likelihood as the probability of a parameter having some value. Above, we define the sig() function as the signum function, which returns the sign of a input.

We define the gradient of the log-likelihood of training pattern v with the following equation:

$$\frac{\partial ln\mathcal{L}(\theta|v)}{\partial w_{i,j}} = -\sum_h p(h|v)\frac{\partial E(v,h)}{\partial w_{i,j}} + \sum_{v,h} p(v,h)\frac{\partial E(v,h)}{\partial w_{i,j}}$$

$$= \sum_{h_i} p(h_i|v)h_i v_j - \sum_v p(v)\sum_h p(h|v)h_i v_j$$

$$= p(H_i = 1|v)v_j - \sum_v p(v)p(H_i = 1|v)v_j$$

The mean of the gradient over training set $S = \{v_1,\ldots,v_\ell\}$ is given as

$$\frac{1}{\ell}\sum_{v \in S}\frac{\partial ln\mathcal{L}(\theta|v)}{\partial w_{i,j}} = \frac{1}{\ell}\sum_{v \in S}\left[-\mathbb{E}_{p(h|v)}\left[\frac{\partial E(v,h)}{\partial w_{i,j}}\right] + \mathbb{E}_{p(h,v)}\left[\frac{\partial E(v,h)}{w_{i,j}}\right]\right]$$

$$= \frac{1}{\ell}\sum_{v \in S}\left[\mathbb{E}_{p(h|v)}\left[v_i h_j\right] - \mathbb{E}_{p(h,v)}\left[v_i h_j\right]\right]$$

$$= \langle v_i h_j \rangle_{p(h|v)q(v)} - \langle v_i h_j \rangle_{p(h,v)}$$

where

$$\sum_{v \in S} \frac{\partial ln\mathcal{L}(\theta|v)}{\partial w_{i,j}} \propto \langle v_i h_j \rangle_{data} - \langle v_i h_j \rangle_{model}$$

$$\frac{\partial ln\mathcal{L}(\theta|v)}{\partial b_j} = v_j - \sum_v p(v) v_j,$$

$$\frac{\partial ln\mathcal{L}(\theta|v)}{\partial c_j} = p(H=1|v) - \sum_v p(v) p(H_i=1|v)$$

Now, returning to our initial discussion, we approximate the gradient of the log-likelihood of training pattern v as the following:

$$CD_k(\theta, v^0) = -\sum_h p(h|v^0) \frac{\partial E(v^0, h)}{\partial \theta} + \sum_h p(h|v^k) \frac{\partial E(v^k, h)}{\partial \theta}$$

The derivatives of each single parameter are calculated from the approximation just given with respect to the expectations over $p(v)$. In *batch learning*, we compute the gradient over the entirety of the training set. However, there are instances where it would be computationally more efficient to run this approximation over a subset of the training data set, which we denote as a *mini-batch*. If we evaluate a single element of the training set when performing this approximation, it's known as *online learning*. In RBMs, we refer to the *reconstruction error* as the difference between the actual input and the predicted input, which falls drastically from the beginning of training moving forward. It is suggested that this metric be used, but proceed with caution. CD learning is approximately optimizing the KL divergence between the training data and the data produced by the RBM and the Gibbs chain's mixing rate. That said, the reconstruction error often can be deceptively small if the mixing rate is also small. As the weights within the RBM increase, typically we observe the mixing rate to move inversely. But a lower mixing rate doesn't always necessarily mean a model is superior to one in which there is a higher mixing rate.

RBM weights, similar to other deep learning models, are typically initialized using values randomly sampled from a normal distribution or other infinitesimally small values. With respect to the learning rate, the same considerations with gradient methods must be taken into account, particularly being careful not to choose a learning rate that's too large or too small. With that being said, an adaptive learning rate may cause issues as it will give the appearance that the model is improving due to a lower reconstruction error, however, as explained earlier, this may not always be the case. It is recommended that each weight update generally be about 10^{-3} times the current weights. Initial hidden biases and weights typically are initialized by selecting them randomly from a normal distribution, as is standard operating procedure for other neural network models.

Momentum Within RBMs

To increase the speed of learning within an RBM, *momentum* is a recommended method. Imagine a gradient plot such as the one in Figure 7-2. If we can imagine the error represented by a point on one of the circles, the dot gains "momentum" as it moves closer to the minimum—but it loses momentum if it tries to go past that point and upwards along the sphere on the opposite side.

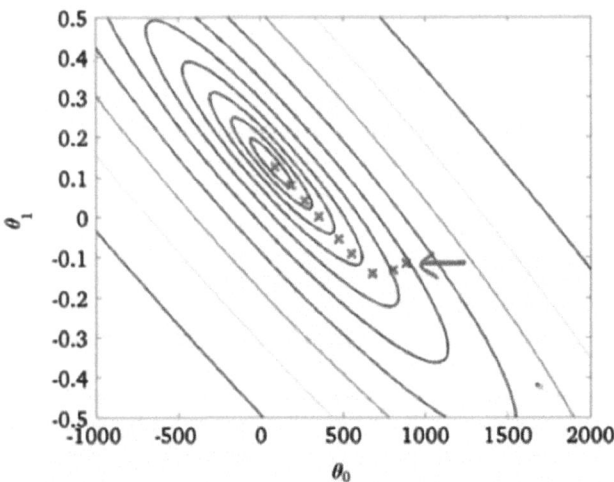

Figure 7-2. Gradient plot

Rather than the traditional gradient descent formula, the momentum method incrementally affects the velocity of the parameter update. We define *momentum* as the percentage of the velocity that is still present after a given epoch; we assume that over time the velocity of a parameter decays. In effect, the momentum method causes the update of the parameters to move in a direction that is not the steepest descent, as with a typical gradient method except less intricate. When using the momentum method, it is suggested that the momentum parameter, α, be set to .5. When it becomes more difficult to reduce the reconstruction error any further, the momentum should be increased to .9. If we notice instability in the reconstruction error—typically noted by occasional, incremental increases—we reduce the learning rate by factors of 2 until this phenomenon subsides. We define the momentum method of updating a parameter as follows:

$$\Delta\theta_i(t) = v_i(t) = \alpha v_i(t-1) - \epsilon \frac{dE(t)}{d\theta_i}$$

Weight Decay

Weight decay can be viewed as a form of regularization, similar to that of the parameter regularization seen in ridge regression and/or LASSO. In RMBs, we typically use a Euclidean norm, which we denote as cost of the weights. Commonly, practitioners take the derivative of the penalty term and multiply it by the learning rate. This prevents the learning rate from changing the objective function we are trying to optimize. Weight decay helps reduce overfitting in such a way that the solution achieved doesn't have units with unusually large weights or weights that are either always on or off. It also improves the mixing rate, in reference to the Gibbs sampling we perform, making CD learning more accurate. Geoffrey Hinton suggests that initially a weight cost of 0.0001 be used.

Sparsity

Generally, a good model is one that has hidden units that are active only part of the time. The reason is that models with sparsely active units are considerably easier to interpret compared to models that are densely populated with active units. We can achieve *sparsity* by specifying the probability of a unit being active, performed by using regularization. This probability is denoted by q and is estimated by

$$q_{new} = \lambda q_{old} + (1-\lambda) q_{current},$$

where $q_{current}$ = mean activiation probability of hidden unit

The natural penalty measure to use is the cross entropy between the desired and actual distributions:

$$\text{Sparsity penalty} \propto -p \log q - (1-p) \log(1-q)$$

As suggested by Hinton, we seek to have a sparsity target as low as 0.1^9 and as high as 0.01. We denote the decay rate as λ, which refers to the estimated sparsity value. This should be no higher than 0.99 but higher than 0.9. We should reduce the sparsity cost if the probabilities we calculate are clustering around the target value, and a general suggestion for modeling this is to collect a histogram of mean activities when collecting random samples.

No. and Type Hidden Units

Being that often the main consideration is that we seek to avoid overfitting. As such, we generally will try to use fewer hidden units rather than more. Particularly, if the data across the observations tends to be very homogenous, we also should try and use fewer rather than more hidden units. However, an instance in which it is reasonable to use more hidden units than normal would be if the sparsity target we're trying to achieve happens to fall within a very small range (or is very small itself). As for the type of units,

CHAPTER 7 ■ AUTOENCODERS, RESTRICTED BOLTZMANN MACHINES, AND DEEP BELIEF NETWORKS

we can use Gaussian visible (and/or hidden), in addition to sigmoid and softmax units denoted by the following, respectively:

$$E(v,h) = \sum_{i \in v} \frac{(v_i - a_i)^2}{2\sigma_i^2} - \sum_{j \in h} b_j h_j - \sum_{i,j} \frac{v_i}{\sigma_i} h_j w_{i,j},$$

$$E(v,h) = \sum_{i \in v} \frac{(v_i - a_i)^2}{2\sigma_i^2} + \sum_{j \in h} \frac{(h_j - b_j)^2}{2\sigma_j^2} - \sum_{i,j} \frac{v_i}{\sigma_i} h_j w_{i,j},$$

$$p = \frac{1}{1 + e^{-x}}$$

$$p_j = \frac{e^{x_j}}{\sum_{i=1}^{K} e^{x_i}}.$$

Deep Belief Networks (DBNs)

The final model I'll address is the deep belief network (DBN), shown in Figure 7-3, another innovation from Geoffrey Hinton. To make a DBN, we stack together restricted Boltzmann machines and train the layers one at a time. Typically, we use DBNs for unsupervised learning problems.

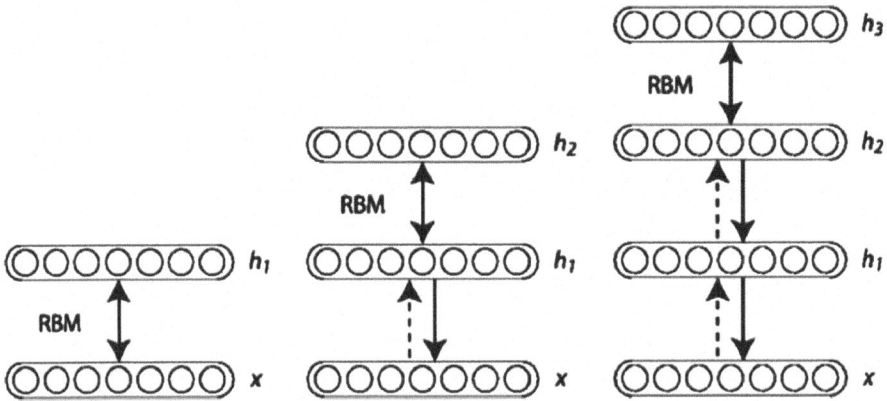

Figure 7-3. Visualization of a deep belief network

In a 2006 paper, Geoffrey Hinton and Simon Osindero, both researchers at the University of Toronto, describe an algorithm useful for fast learning. The difficulty posed by training networks with many hidden layers inspired the creation of a hybrid model. The main attraction of this model, in relation to the training problem, is that by design there

CHAPTER 7 ■ AUTOENCODERS, RESTRICTED BOLTZMANN MACHINES, AND DEEP BELIEF NETWORKS

are complementary priors that allow us to easily draw from the conditional probability distribution. This is done by starting with a random configuration a layer deep within the network. We then pass through each layer of the network, in which the state of a given layer is determined by a Bernoulli trial. The parameters for the Bernoulli function are derived from the input received from the preceding layer in the initial "top-down" pass.

Fast Learning Algorithm (Hinton and Osindero 2006)

Data is generated from an RBM by taking a random state within a given layer and performing Gibbs sampling over it. Simply stated, *Gibbs sampling* is a type of Monte Carlo method in which we try to obtain a sequence based on a probability distribution that the user specifies, but which the algorithm tries to approximate. Typically, the distribution is multivariate. All units within a chosen layer are updated in a parallel fashion, and this is repeated until we've determined to be sampling from the equilibrium distribution. In Figure 7-4, we can see the visible and the hidden layers of an RBM.

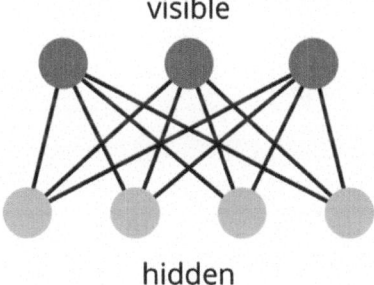

Figure 7-4. *Visualization of restricted Boltzmann machine*

Each weight uses a visible unit, *i*, and a hidden unit, *j*. When a data vector is "clamped" on the visible units, the hidden units are sampled from their conditional distribution, which is factorial. The gradient of the log probability is given by the following:

$$\frac{\partial \log p(v^0)}{\partial w_{i,j}} = \langle v_i^0 h_j^0 \rangle - \langle v_i^\infty h_j^\infty \rangle$$

When we minimize the KL divergence, we in effect maximize the log probability. If you would like to learn complicated models, break up the single model into smaller, simpler models. After this point, these models can be learned sequentially. An example of this sequential learning would be gradient boosting, as discussed in Chapter 3. Reasonable approximations for W_0 are learned based on the assumption that higher layers derive the complimentary prior for W_0. In practice, we can achieve this outcome by assuming that all the weight matrices must be equal to one another. When solving this constrained optimization problem, learning becomes significantly easier than before, and the problem itself is reduced to learning an RBM, whereupon good approximate solutions are achieved via minimizing contrastive divergence.

Algorithm Steps

1. Under the assumption that all the weight matrices are tied, learn $\mathbf{W}_0$.

2. Use $\mathbf{W}_0^T$ to infer factorial approximate posterior distributions over the states of the variables in the first hidden layer.

3. Learn an RBM model with respect to high-level abstractions of the data generated by $\mathbf{W}_0^T$.

4. Repeat until convergence upon an optimal solution.

If the weight matrices in the higher levels of the model change, we are guaranteed to see improvements in the model. The bound given becomes an equality if $Q(.|\mathbf{v}^0)$ is the true posterior of the data. Hinton specifically suggests a greedy learning method, as described in Neal and Hinton (1998). The energy of a given configuration of $\mathbf{v}^0, \mathbf{h}^0$ is defined as

$$E(v^0, h^0) = -\left[\log p(h^0) + \log p(v^0 | h^0)\right],$$

with a bound of

$$\log p(v^0) \geq \sum_{all\ h^0} Q(h^0 | v^0)\left[\log p(h^0) + \log(v^0 | h^0)\right] - \sum_{all\ h^0} Q(h^0 | v^0) \log Q(h^0 | v^0)$$

where $\mathbf{h}^0$ = binary configuration the initial hidden layer units, $p(\mathbf{h}^0)$ = the prior of the current model $\mathbf{h}^0$, and $Q(.|\mathbf{v}^0)$ = probability distribution over the initial hidden layer's binary configurations.

Summary

This brings us to the end of discussing autoencoders, RBMs, and DBNs. This also concludes all the chapters on deep learning models. Now that we've discussed these models, it's time to turn our attention to experimental design and feature selection techniques to help you increase the accuracy of your machine learning models.

CHAPTER 8

■ ■ ■

Experimental Design and Heuristics

After having reviewed all the machine learning and deep learning models that will be relevant to problem solving that you will encounter, it's finally time to talk about useful methods of structuring your research, both formal and informal.

Beyond just knowing how to properly evaluate the solutions developed, you should be familiar with the concepts associated with the field of *experimental design*. Ronald Fisher, an English statistician prominent in the 20th century, was one of the most influential figures in the field of statistics. His techniques are frequently referenced when performing experimentation and are useful to review even if you don't use them explicitly.

Analysis of Variance (ANOVA)

ANOVA is group of methods that are used to study the variation among groups of observations within data. An extension of the z and t test, and similar to regression, we observe the interaction between the response and explanatory variables. We assume that the observations within the data are independent and identically distributed (IID) normal random variables, that residuals are normally distributed, and that variance is homogenous. Among the multiple ANOVA models are the following ones discussed in the rest of this section.

One-Way ANOVA

Used to compare three or more sample spaces' means/averages to one another. Specifically, it's used in cases where the classification is performed by one variable/factor that has two or more levels.

Two-Way (Multiple-Way) ANOVA

This is similar to one-way ANOVA, except this model can be used where there are two or more explanatory variables.

Mixed-Design ANOVA

In contrast to the prior models described, mixed-design ANOVA is distinguished by having one of the factor variables be analyzed across subjects and the other factor be a within-subjects variable.

Multivariate ANOVA (MANOVA)

This one is similar to one-way and two-way ANOVA, except it is particularly used to analyze multivariate sample means, or when there are two or more explanatory variables in a given data set.

Having addressed the various ANOVA models, the next section talks about the method by which we evaluate the results: the F-statistic.

F-Statistic and F-Distribution

Named after Ronald Fisher, the F-statistic is the ratio of two statistical variances. F-statistics are based upon the F-distribution, a continuous probability distribution (see Figure 8-1). We denote this distribution as the null distribution of a given test statistic for the F-test. Let's assume we have variables A and B such that they both have chi-square distributions with n and d degrees of freedom respectively such that

$$X = \frac{\frac{A}{n}}{\frac{V}{d}},$$

$$f(x) = \left(\frac{\Gamma\left(\frac{n}{2} + \frac{d}{2}\right)}{\Gamma\left(\frac{n}{2}\right)\Gamma\left(\frac{d}{2}\right)} \right) \left(\frac{n}{d}\right) \frac{\left[\left(\frac{n}{d}\right)x\right]^{\frac{n}{2}-1}}{\left[1 + \left(\frac{n}{d}\right)x\right]^{\frac{n}{2}+\frac{d}{2}}}, \quad x \in (0, \infty)$$

CHAPTER 8 ■ EXPERIMENTAL DESIGN AND HEURISTICS

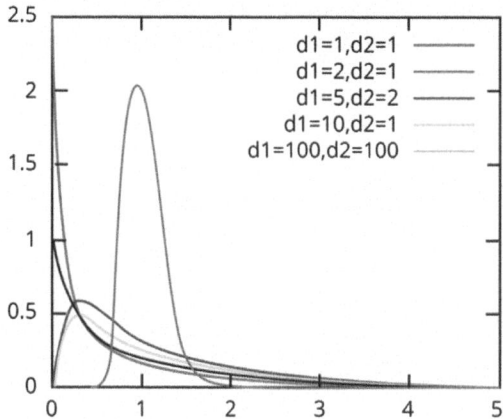

Figure 8-1. *PDF for F-distribution*

Say, for example, we're considering a one-way ANOVA and assume that the means of a set of populations are equal and normally distributed. We define the F-statistic as

$$F = \frac{\frac{SSE}{k}}{\frac{SSR}{n-k-1}} = \left(\frac{\frac{\sum(\hat{Y}_i - \bar{Y})^2}{k}}{\left(\frac{\sum(Y_i - \hat{Y}_i)^2}{n-k-1}\right)} \right),$$

where k is degrees of freedom and n is the number of n response variables. The null hypothesis states that a model created using only an x-intercept and a model created by the user yield indistinguishable results (within a given confidence interval). The alternative hypothesis states that the model the reader creates is significantly better than a model featuring only the x intercept. Just as when testing any other measure of statistical significance, this is determined based on the threshold we want to set. (90% level of confidence, 95% level of confidence, and so on).

Let's now use a toy example to apply and explain the concepts we just have addressed. For this example, we will be using the iris data set:

```
#Loading Data
data("iris")

#Simple ANOVA
#Toy Example Using Iris Data as Y
y <- iris[, 1]
x <- seq(1, length(y), 1)
plot(y)
```

CHAPTER 8 ■ EXPERIMENTAL DESIGN AND HEURISTICS

The data set will be utilized to create response and/or explanatory variables in the following experiments. In the first toy example, we take the first column of the iris data set (representing the sepal length of each observation) and make this explanatory variable. However, before we perform a one-way ANOVA, let's validate the assumptions necessary to fit data to a linear model. We'll begin by visually inspecting our data, as shown in Figure 8-2.

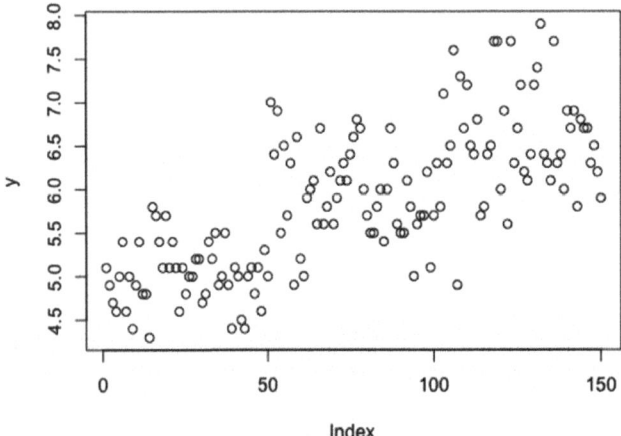

Figure 8-2. Visualization of data

Immediately, we notice that the data is fairly linear in its orientation, featuring a positive slope. This is a good first indicator, but we should dig deeper to ensure that the rest of our assumptions are satisfied. In this instance, we'll focus on plotting the residuals of a fitted model. By *residuals*, I mean the quantity left over from the remainder of the actual value minus the value predicted by the model. You should heavily utilize residual analysis when working with linear models, but also in general, because they provide great visual insight into how well a particular model works as well as the orientation of the data. In Figure 8-3, we see the following plots created from fitting a linear model for x and y:

plot(glm(y~x))

CHAPTER 8 ■ EXPERIMENTAL DESIGN AND HEURISTICS

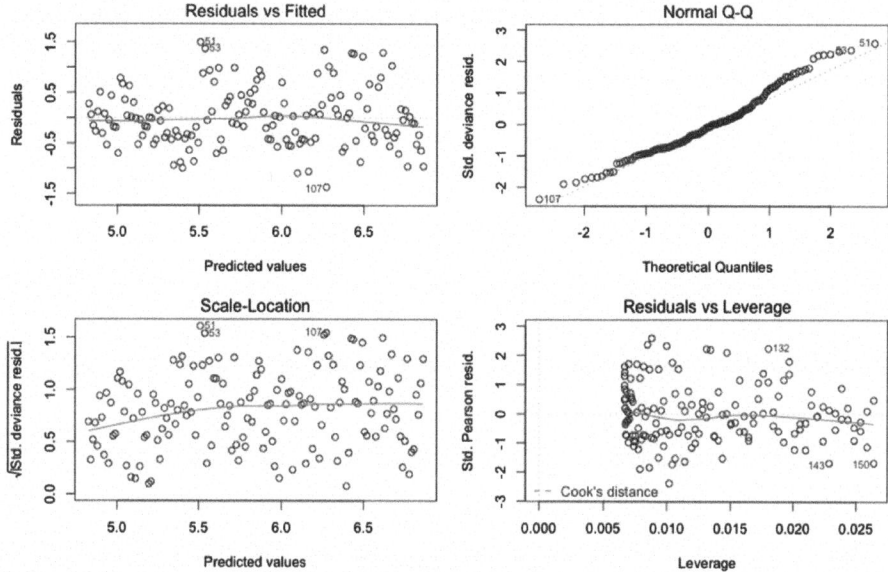

Figure 8-3. Residual plot

Note the graph in the top righthand corner of the four displayed in Figure 8-3. This is a *quantile plot*, which effectively displays how much the distribution of the residuals is normal. When closely inspecting the plot, we can see that a considerable amount of the data lies on the dotted 45-degree angle line, which is the marker of normality in the data. However—and as is the case usually—we notice that the tail ends tend to lift slightly above this line. It's useful to note that almost always, the data we will qualify as being normally distributed will exhibit similar patterns. In the real world, most data tends to be close to normally distributed when we have enough of it, but it's unlikely that it will be perfectly normally distributed. As such, we accept here that the data is normally distributed and move onto validating the remaining assumptions. When the data is normally distributed, it can be fit to a linear model and therefore we can reasonably estimate the values within the range of the x variable.

Because we also require that errors exhibit constant variance, let's turn our attention to the plot in the top lefthand corner. Note here a plot with an x-axis that denotes the value that the regression outputted and a y-axis detailing the value of the residual. The horizontal line through the center of the plot represents the region where the fitted value is equal to the actual value, or where the residual for an observation is zero. When referring specifically to our data, we can see that generally speaking, the shape of the residuals plotted seems to be consistent from the left to the right side of the plot. As such, we would state that the residuals in fact do exhibit constant variance. If not, we would notice that there would distinct patterns in the shape of the scatter plot that would either become more exaggerated or less exaggerated from the left to right side of the plot.

141

CHAPTER 8 ■ EXPERIMENTAL DESIGN AND HEURISTICS

Most importantly, pay attention to the plot on the bottom righthand side of the figure. It addresses an important concept for understanding how certain data points can alter the fitted line of the regression model. *Leverage* is described as a relative measure to how large the difference in value a particular observation is from the rest of the data set. Observations that specifically have high leverage are denoted in R by placing the index adjacent to the data point. We define leverage with

$$h_i = \frac{X_i - \bar{X}}{\sum_{j=1}^{n}(X_j - \bar{X})} + \frac{1}{n}$$

where n = number of observations, X_i = ith observation of X, $\bar{X}$ = mean of all the observations within X, and $i = 1,2,...,n$.

Highly related to leverage is the concept of *Cook's distance*, which directly estimates the influence a specific observation has on this regression model. We define Cook's distance as

$$D_i = \frac{e_i^2}{s^2 p}\left[\frac{h_i}{(1-h_i)^2}\right],$$

where e_i^2 = squared residuals of a given observation, s^2 = mean squared error of the model, p = the number of parameters in a model, h_i = ith diagonal of the H matrix where $H = X(X^T X)^{-1} X^T y$, $i = 1,2,...,n$, and n = number of observations.

Typically, we consider an observation as being particularly influential if its Cook's distance value is greater than 1 or if its distance value is greater than 4/n. Which threshold to use is ultimately up to you, but it's obvious that this will depend on the case, and it's worth inspecting on an experimental basis which provides a data set with more or fewer outliers, and how that would affect your end goal. If, for example, the purpose of an experiment is anomaly detection, it might be foolish to reduce the threshold such that more noise in the data set is qualified as a signal. When referring back to our specific plot, we can see that a considerable amount of data points are being flagged as being influential. We will keep this in mind as we move forward with our model choice.

When assessing all the plots in the data set, we can confidently say that although there are outliers, and our assumptions aren't met perfectly, the robustness of OLS regression allows these slight deviations to be overcome. As such, it's reasonable to choose OLS regression as a model for this task, and therefore ANOVA will yield statistically significant results. When executing the code, we observe the following:

```
simpleAOV <- aov(y ~ x)
summary(simpleAOV)
```

```
            Df  Sum Sq  Mean Sq  F value  Pr(>F)
x            1   52.48    52.48    156.3  <2e-16 ***
Residuals  148   49.69     0.34
```

Just as when we use the `summary()` function on a glm object, we are given a measure of its statistical significance. Instead of a Z-score, though, we're given an F-score from the concept addressed prior to this example and its relative p score. In this instance, we can say with greater than 99% significance that the results we reject the null hypothesis. As such, this model is a significantly better fit than an intercept-only model, and therefore we can be more confident in its results. However, let's say we'd like to compare more than one fitted model. As such, let's inspect what happens when we include more than one variable, but study the interaction between the two of them as well.

As we can see in the following code, we use the second and third columns as explanatory variables in this model. When fitting our model, we multiply both explanatory variables together. When executing the code, we observe the following results:

```
#Mixed Design Anova
x1 <- iris[,2]
x2 <- iris[,3]
mixedAOV <- aov(y ~ x1*x2)
summary(mixedAOV)

            Df Sum Sq Mean Sq F value   Pr(>F)
x1           1   1.41    1.41   12.9 0.000447 ***
x2           1  84.43   84.43  771.4  < 2e-16 ***
x1:x2        1   0.35    0.35    3.2 0.075712 .
Residuals  146  15.98    0.11
```

Our residuals as significantly smaller, and all of variables are statistically significant within at least a 90% confidence interval. Let's execute the following code and visually compare the two models in Figure 8-4:

```
par(mfrow = c(2,2))
plot(glm(y ~ x1*x2))
dev.off()
```

CHAPTER 8 ■ EXPERIMENTAL DESIGN AND HEURISTICS

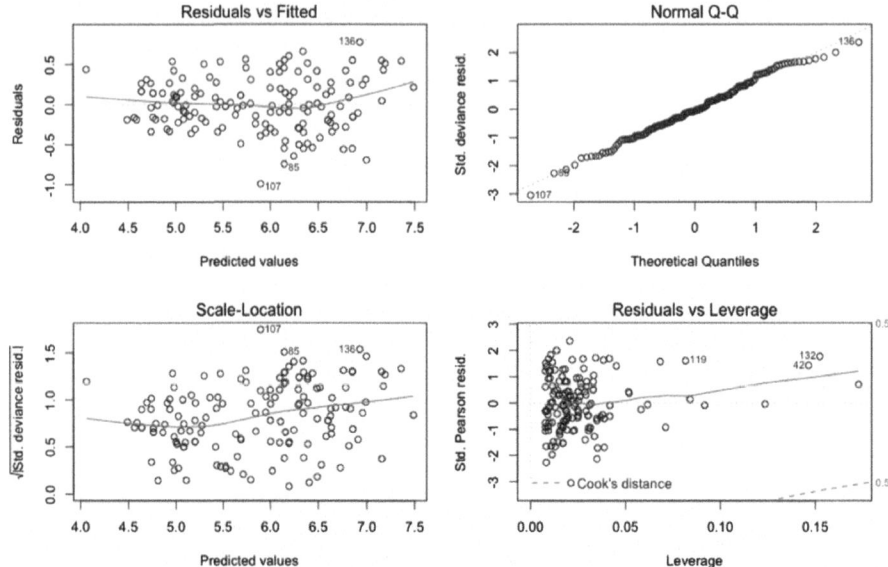

Figure 8-4. *Mixed design ANOVA plot*

We can see that all of the assumptions we need to fulfill are done so significantly better. Virtually all the residuals are normally distributed as displayed in the normal Q-Q plot, the residuals exhibit constant variance, and a considerably smaller amount have leverage. As such, when choosing between the two models we've defined, it's reasonable for us to choose the second of the two in comparison to the first.

This is a brief example of how we can use ANOVA in the course of model selection. In Chapters 10 and 11, you will learn to effectively perform these same analyses with respect to comparing deep learning and machine learning algorithms.

Let's now discuss in greater detail how to structure our experiments, with the guidance of Fisher's principles.

Fisher's Principles

Ronald Fisher, one of the most distinguished statisticians of all time, gave an explanation of principles for experimental design. The following are descriptions of his principles, as well as general advice with respect to how you might want to implement them:

1. *Statement of Experiment*: You should explicitly state the scenario that inspired the experiment, very explicitly giving an outline of the steps that will be taken place in the experiment on a very high level. It is generally accepted that the introduction should include a high-level overview of the topic, and each section should describe a different component in greater detail, logically progressing from beginning of the experiment to the end.

2. *Interpretation and its Reasoned Basis*: From the beginning, it's reasonable to give what you might expect to be the reasonable outcomes. You should state the outcomes that you feel must be considered, but realize that providing an endless list of outcomes for those to whom you report is not likely to be very helpful. Moreover, when discussing all the possible outcomes, do so in a manner that provides actionable insights for those reading your research. Research that does give actionable insights self-evidently leaves more room for misapplication.

3. *The Test of Significance*: In the context of evaluating machine learning and deep learning solutions, a simple suggestion is to bootstrap the test statistics used to evaluate a given model. It's reasonable to assume that if you draw enough test statistics over a long enough time, the data will be normally distributed. From this point, a Z-test can be performed to determine the reasonable level of statistical confidence one has in the model.

4. *The Null Hypothesis*: This hypothesis should state that the results shown have no significance, and any deviation between testing populations is due to some extraneous error such as improper sampling or deviations from proper experimental practices. This must be a component of all statistical testing.

5. *Randomization: The Physical Basis of the Validity of the Test*: When performing a test, the results reached should be performed in a manner such that this outcome was not biased. In some cases, this may require randomized observations of data to remove any inherent biases present in the modeling of the experiment that would lead to some results.

6. *Statistical Replication*: The results reached from a test should and must be replicable. Results reached that are unreasonable given the constraints inherent to the data set and environment in which we expect to observe such an occurrence are not as valuable as results that are replicable.

7. *Blocking*: The process by which different experimental groups are compartmentalized such that different variations and biases are reduced or prevented entirely from affecting the results of an experiment.

Plackett-Burman Designs

Created in the 1940s by Robin Plackett and J. P. Burman, Plackett-Burman designs are a method of finding a quantifiable dependence of explanatory variables, which we call *factors* in this case, of which each factor has L levels. The overall objective is to minimize the variance of the estimates of dependencies using a limited amount of experiments. To fulfill this goal, an experimental design is chosen such that each combination for any given pair of factors appears an equal number of times throughout each experimental "run."

The Plackett-Burman design requires a small number of experiments, specifically a multiple of 4 up to 36, and that the design have N samples that can study up to k parameters, where k = N − 1. In the case that L = 2, an orthogonal matrix in which each element is either −1 or 1 is used. This matrix is also known as a *Hadamard matrix*. This method is useful for identifying the main effects of different factors on the response variable, such that we can eliminate factors that seem to have little to no effect. Plackett and Burman themselves give specific designs for L equaling 3, 4, 5, and 7.

Have a look at the matrix in Figure 8-5, which visually depicts a Plackett-Burman design. When performing this *design of experiments* (DOE), you must write the appropriate row as the first row of the design table. In this instance, we begin with a +, −, +, −, +, +. This is a permutation of the sequence that appears in every row, which represents a treatment combination. You can think of a treatment combination as a unique combination of a feature set. The second row is then created by shifting sequence in the prior row to the right by one column. This process is repeated for each of the remaining rows. The final row features all negative elements. It's important to recognize, though, that Plackett-Burman designs can't describe whether the effect on a given factor will result in the effect of another, and it similarly can't know the effects themselves given a small enough design. This design is considered to be a preparatory step to data analysis, and it's suggested that alternative preparatory steps be juxtaposed alongside it in addition to other steps taken subsequently afterwards.

Plackett-Burman 8-Run Matrix

				Factors				
		A	B	C	D	E	F	G
Treatment Combinations	1	+	−	−	+	−	+	+
	2	+	+	−	−	+	−	+
	3	+	+	+	−	−	+	−
	4	−	+	+	+	−	−	+
	5	+	−	+	+	+	−	−
	6	−	+	−	+	+	+	−
	7	−	−	+	−	+	+	+
	8	−	−	−	−	−	−	−

Figure 8-5. *Plackett-Burman matrix*

Space Filling

These methods don't require discrete parameters, and the sample size is chosen independently from the total number of parameters. These are recommended for instances in which the reader would like to create response surfaces, but it should be noted that it becomes difficult to determine the main effects and interactions of a given or set of parameters respectively.

Full Factorial

Full factorial is one of the most popular methods of experimental design, in which $N = 2\wedge K$, where k is equal to the number of factors. As an example, let's have k factors where $L = 2$. In this model, we don't distinguish between nuisance and primary factors prior to the experiment taking place. Given that $L = 2$, we will denote them as a high, "h", or low, "l", level. High-level factors receive a value of 1, and low-level factors receive a value of -1. We determine the interaction of the variables as the product of the individual factors. From any experiment that is possible given the factorial constraint, the samples in which the factors are changed one at a time are still a part of the sample space. This allows for the effect of each factor over the response variable. Let's now also define M as the main intersection of a variable X. This is the difference between the average response variable at the high-level samples and the average response at the low-level samples. If we have three factors with two levels per factor, M for X_1 would be defined as the following:

$$M_{X_1} = \frac{y_{h,h,h} + y_{h,h,l} + y_{h,l,h} + y_{h,l,l}}{4} - \frac{y_{l,l,l} + y_{l,l,h} + y_{l,h,l} + y_{l,h,h}}{4}$$

If we wanted to see the interaction between two or more factors, the equation would be the same, except the interaction of the variables would be represented by the product of the variables, rather than the individual values a factor possesses at a given state. Both of the main effects and the intersection effect statistics give an effective method of determining the degree to which individual, or combinations of, factors affect the response variable. Full factorial designs do not complicate the data in any such manner and present a transparent method of examining variable effects. If there are more than two levels, an adjustment must be made to take the average effect of all the levels on a given response variable, where the denominator is N, such that

$$\bar{y} = \frac{\sum_i^{L_1} \sum_j^{L_2} \sum_l^{L_3} \sum_m^{L_4} y_{i,j,l,m}}{N}$$

Halton, Faure, and Sobol Sequences

Within the umbrella of space filling techniques, many of these are motivated by pseudo-random number generators. *Pseudo-random numbers* are series-generating sets that pass randomness tests. We denote a pseudo-random number generator as the following function:

$$\phi:[0,1) \to [0,1), \quad \gamma_k = \phi(\gamma_{k-1}), \quad k=1,2,\ldots$$

We must choose a value of ϕ that gives a uniform distribution of γ_k. A popular method to achieve this is the Van der Corput sequence, where we have a base, b, ≥ 2 and successive integer numbers n are expressed in their b-adic expansion form such that the following is true

$$n = \sum_{j=1}^{T} a_j b^{j-1},$$

$$\varphi_b : N_0 \to [0,1),$$

$$\varphi_b(n) = \sum_{j=1}^{T} \frac{a_j}{b^j}$$

where a represents the coefficients of the expansion.

Halton sequences use base-two, base-three, and base-five for Van der Corput sequences in first, second, and third dimensions respectively. This pattern continues such that prime numbers are used for the base in every successive dimension. With this being said, multidimensional clustering causes high correlations between dimensions, effectively defeating the purpose of experimental design in and of itself. In an effort to combat this problem, Faure and Sobol sequences use only one base for all dimensions and a different permutation of the vector elements for each dimension.

A/B Testing

When designing applications, websites, and/or dashboard applications, it's useful to determine the effect changes in certain functionality have on the product. We can imagine, for example, that an engineer is trying to determine with some statistical certainty whether the implementation of a new feature has had an effect on acquiring new users. For such situations, it's recommended that the person use something known as A/B testing. Broadly, *A/B testing* refers to the statistical hypothesis testing methods used to compare two data sets, a control group and a test group, which are A and B respectively. We can also modify the test such that we can test A and multiple additional control tests.

The motivation for A/B testing is simple in that the development of different products, regardless of whether they have machine learning or deep learning capabilities, allows us to determine with statistical confidence whether we have made improvements from the original iteration to the next. That said, we can use these processes as a series of experimentations to iteratively move from one generation of software to the next to observe the improvements in efficiency. Commonly, the beta-binomial hierarchical model is one of the most popular methods by which we can A/B test a control group over a multiple test groups. As such, we will review this model. First, however, let's review a simple two-sample A/B test.

Simple Two-Sample A/B Test

Assume that here we're comparing one control group against one test group and that we're trying to see whether our new website generates more clicks due to feature changes. We will firmly show that although this test is stable for two examples, you should avoid using this for more than two samples. Let's say that we have two data sets representing the different attributes of the various websites and we want to test within a 95% confidence level. For this, we would use a t-test. Now let's also assume after the t-test is performed, we observe that the difference in the means is significantly different and that x2 is significantly improved from the prior model. Now let's assume that we keep on making different versions of our web page and continuously try to use this model. After nine different tests, x2 still is proving to be the most superior model. But when we run x2, we actually see no difference in improvement from clicks from x2 to the other websites. This common problem with two-sample A/B testing is due to false positives.

Next, I'll show the probability of ten individual hypothesis tests showing correct results via the binomial distribution. Let event A = x2 let's say better than nine other counterparts at 90% confidence interval, B = x2 is better than nine other counterparts at 95% confidence interval, and C = x2 better than nine other counterparts at 99% confidence interval:

$$P(A) = .90^{10} = 34.87\%, \quad P(B) = .95^{10} = 59.87\%, \quad P(C) = .99^{10} = 90.44\%$$

Stated simply, under events A, B, and C we could expect that our experiments yield 6.5, 4.013, and .95 false positives. Although the 99% confidence interval performs the best in this example, we can see under the other confidence intervals why this methodology would become a problem. As such, for testing multiple groups, it is recommended that we use the beta-binomial distribution.

Beta-Binomial Hierarchical Model for A/B Testing

Bayesian statistics is a school of thought on the concept of probability. Here, it becomes the theoretical underpinning for this model and also can be used to provide modified hierarchical models with respect to the distribution. In Bayesian statistics, we often refer to the prior and posterior distributions. The *prior* distribution refers to the probability distribution with respect to some parameter (data that we have already acquired), whereas the *posterior* refers to the probability distribution of some parameter with respect to the data (data which we want to acquire). The prior distribution and the

CHAPTER 8 ■ EXPERIMENTAL DESIGN AND HEURISTICS

posterior distribution form a *conjugate* distribution. For ease of analysis, we typically seek to use distributions within the same family for the respect prior and posterior distributions, and that's why in this hierarchical model we are using the beta and binomial distributions.

The *beta* distribution is a probability distribution bound within the interval [0,1] with parameters α and β that ultimately control the shape of the distribution. Typically, we use the beta distribution to statistically model random variables. As stated earlier, within the same family as the beta distribution is the *binomial* distribution. This is often used to model probability distributions that feature independent binary outcomes, such as coin flips. We define the probability density function for both the beta and binomial distributions (see Figures 8-6 and 8-7) respectively as

$$\frac{x^{\alpha-1}(1-x)^{\beta-1}}{\frac{\Gamma(\alpha)\Gamma(\beta)}{\Gamma(\alpha+\beta)}},$$

$$\binom{n}{k}p^k(1-p)^{n-k}$$

where n = the number of successes, k = total number of trials, and p = the probability of success.

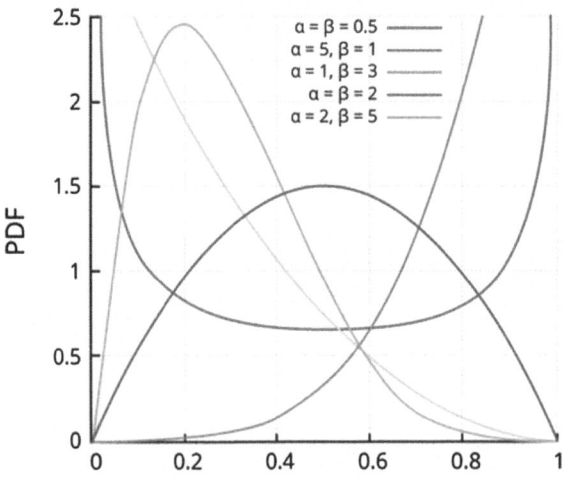

Figure 8-6. Beta distribution

CHAPTER 8 ■ EXPERIMENTAL DESIGN AND HEURISTICS

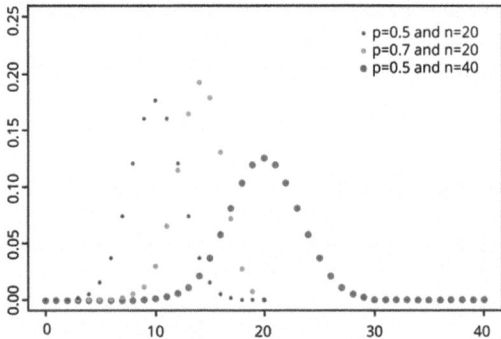

Figure 8-7. Binomial distribution

We then model our posterior expectations from the beta distribution and the prior distribution as the binomial, whereupon we compare the difference in means between the prior and the posterior distributions to compare website performance.

Feature/Variable Selection Techniques

Now that we've discussed several experimental design models, let's talk about steps you would want to take after you have more of an understanding of the factors in a given data set. Variable selection seems to be directly related to experimental design, but this section will discuss more specific algorithms to be used for the purpose of reducing dimensionality and less exploratory methods for analyzing variables and their interactions with the response variable. This is important for a multitude of reasons, but can often be a major element of optimizing machine learning algorithms when deploying them. Feature selection is a much less tedious process than parameter tuning, particularly in deep learning models. As such, it can be a quick way to creating models that train quicker and produce more accurate outputs. As with many techniques described earlier, caution must be taken because too much feature selection can result in creating overfitted models.

Backwards and Forward Selection

Backward selection is one of the simplest variable selection methods and is particularly common when using simple or multiple linear regression. Preliminarily, you should take the data set with all explanatory variables and regress them against the response variable. After this step, choose a statistical significance level appropriate for the given situation (85%, 90%, 95%, and so on). One variable at a time, we remove the variables with the lowest statistical significance from the data set (such as the statistical significance yielded from the summary() function when using a glm() model). We the regress the new subset of the original data set and continue until all the variables in the data set are statistically significant. In forward selection, the process is the same as the prior method, except the distinction is that you start with a model with no variables, add variables, and check their statistical significance. If they're at or above the threshold, they should be added. If not, they should be removed. Considerations to keep in mind when using these methods are to reduce statistical noise considerably but to avoid a model that's overfitted to the test data, particularly if out-of-sample prediction is the end goal to the model being built. You should also be careful to not remove too many variables as to reduce performance.

For deep learning, some models have feature selection embedded into them. Specifically, certain layers within CNNs arguably exist for the purpose of eliminating noise such that the data left is rich with information. Specifically, pooling layers can be thought of as doing this. By reducing the input size, we ease the computational load from input to output while also assisting the algorithm in more accurately tuning the weights between these layers and ultimately classifying an image.

Beyond using P-values, you can choose other statistical criteria to determine which variables to retain/remove. Among those most common are Akaike information criterion (AIC) and Bayes information criterion (BIC):

$$AIC = 2k - 2\ln(\hat{L}),$$

$$BIC = \ln(n)k - 2\ln(\hat{L}),$$

$$\hat{L} = p(x|\hat{\theta}, M)$$

where L is the max likelihood function for the model, $\hat{\theta}$ are the parameters, and k is the number of parameters.

AIC and BIC are very closely related. AIC is based within the field of information theory, and the goal is to choose a model with the minimum possible AIC value. By definition of the function, the greater the magnitude of the log-likelihood, the smaller the AIC value. Henceforth, models that are more closely fit to the data will ultimately have lower AIC values. BIC is ultimately motivated by Bayesian statistics and is similar to AIC. BIC scores specifically are used to evaluate the performance of a model on a training set, where we choose the model that yields the smallest BIC. In particular, BIC penalizes models that have more parameters rather than less. Because of this, BIC inherently prefers models that do not overfit to the data set, hence making a criterion by which you're encouraged to choose a model that generalizes to the data you're analyzing. Note that the BIC can't handle complex collections of models, and it should only be assumed to be valid in instances of when n is substantially greater than k. With respect to considerations for AIC, the AIC values computed must be across the same data. Specifically, it's not an objective measure such as the coefficient of determination.

Principal Component Analysis (PCA)

Principal component analysis (PCA) is one of the most commonly used variable selection techniques that can exclusively be used for numerical data. Mentioned earlier in several examples, PCA is a statistical method used to reduce dimensionality of data sets. Simply stated, we transform the data into new variables called *principal components* and eliminate the principal components that explain negligible amounts of the variance exhibited within the data set. The benefit of this technique is that we preserve the variance of the data set while being able to perform visual and exploratory analysis much easier than prior to the transformation.

Our goal is to find the linear function of random variables from the x vector with the vector of constants from the alpha vector with the maximum variance. This linear function produces our principal components. Be that as it may, each principal component must be in order of decreasing variance, and each principal component must be uncorrelated with each other. Our objective is the following:

$$\text{Maximize } Var(\alpha'_k x) = \alpha'_k \Sigma \, alpha_k$$

We seek to use constrained optimization, because without a constraint the value of a_k could be infinitely large. As such, we'll choose the following normalization constraint, where $\alpha'_k a_k = 1$.

The Lagrange multiplier method is a tool for constrained optimization of differentiable functions. In particular, it's helpful for finding local maxima and minima of a respective function subject to a given constraint. Within the context of the experiment, the Lagrange multipliers are applied as follows

$$\alpha'_k \Sigma \, a_k - \lambda(\alpha'_k a_k - 1),$$

$$\frac{d(\alpha'_k \Sigma \, a_k - \lambda(\alpha'_k a_k - 1))}{d\alpha_k} = 0,$$

$$\Sigma \, \alpha_k - \lambda \alpha_k = 0,$$

$$\Sigma \, \alpha_k = \lambda_k \alpha_k$$

with the final step of the equation yielding the eigenvector α_k and its corresponding eigenvalue λ_k. Our objective is to maximize λ_k, and with the eigenvectors defined in decreasing order. If λ_1 is the largest eigenvector, then the first principal component is defined as $\Sigma \, \alpha_1 = \lambda \alpha_1$. In general, we define a given eigenvector as the k-th principal component of x and that the variance of a given eigenvector is denoted by its corresponding eigenvalue. I'll now demonstrate this process when k = 1 and when k > 2. The second principal component maximizes the variance subject to being uncorrelated with the first principal component with the non-correlation constraint being as follows:

$$cov(\alpha'_1 x \alpha'_2 x) = \alpha'_1 \Sigma \, \alpha_2 = \alpha'_2 \Sigma \, \alpha_1 = \alpha'_2 \lambda_1 \alpha'_1 = \lambda_1 \alpha'_2 \alpha_2 = 0,$$

$$\alpha'_2 \Sigma \, \alpha_2 - \lambda_2(\alpha'_2 \alpha_2 - 1) - \phi \alpha'_2 \alpha_1$$

$$\frac{d\left(\alpha_2'\sum\alpha_2 - \lambda_2\left(\alpha_2'\alpha_2 - 1\right) - \phi\alpha_2'\alpha_1\right)}{d\alpha_2} = \sum\alpha_2 - \lambda_2\alpha_2 - \phi\alpha_1 = 0,$$

$$\alpha_1'\sum\alpha_2 - \alpha_1'\lambda_2\alpha_2 - \alpha_1'\phi\alpha_1 = 0,$$

$$0 - 0 - \phi 1 = 0,$$

$$\phi = 0,$$

$$\sum\alpha_2 - \lambda_2\alpha_2 = 0$$

This process can be repeated up to k = p, yielding principal components for each of the p random variables. Limitations associated with PCA are numerous, though, and must be considered for the problem type. Foremost, PCA assumes that there are linear correlations across features. Obviously, this is not necessarily always the case in a practical context and therefore renders the results yielded by PCA questionable. Secondly, PCA only can be used on numerical data sets and the downfalls of numerically encoding categorical data (discussed later in this chapter) can add implicit biases that render the results of this technique useless. Moreover, PCA explicitly assumes variance is the most important statistic with regard to analyzing a data set. Although variance is often an important statistic, in some problem cases it might not necessarily be.

An example of how PCA can be applied to deep learning is through the process of PCA whitening. When we refer to *whitening*, we mean the process of making the input data less homogenous, in an effort to make the data less homogenous from one observation to another. In the instance of a CNN, this can be of great use for image classification. Specifically, in image data many pixels adjacent to one another often have similar, if not the same, values within a large region.

An example of this would be to look at the MNIST data set and see which patches of the image are black versus which are white. PCA whitening instead yields an eigendecomposition of the matrix such that this homogeneity is removed. As such, the features of each individual are significantly less similar than in their original form, but the variance within the data is preserved, as is a benefit when performing an eigendecomposition on a matrix.

Factor Analysis

Factors are unobservable variables that are highly correlated with one another and that influence a given explanatory variable. Unlike the ultimate purpose of PCA, dimensionality reduction, *factor analysis* seeks to locate independent variables. Moreover, we would like to determine what influence the factors have on the surface attributes. It's built from the assumption that observed variables can be reduced to a subset which exhibit similar variance. In factor analysis, we require that the data must be normally distributed and that there are virtually no outliers within the data set. We also should seek to analyze data that is numerous in it's observations, and the correlations, while not nearly linear as to avoid multicollinearity, must be moderate to high across the data set. The typical factor analysis model is given by

$$X_j = a_{j1}F_1 + a_{j2}F_2 + \ldots + a_{jm}F_m + e_j, \, j = 1, \ldots, p$$

Where e_j = the unique and specific factor to a given explanatory variable, j = factor loadings, X_j = an explanatory variable, and m = the underlying factors

Factor loadings can be thought of as weights, where they denote the degree to which they influence a given factor with respect to an individual variable. Surface attributes are denoted as the individual explanatory variables. Typically, a factor analysis model will yield factors such that there are no correlations between the individual variables, so we have independent variables, similar to principal components. It should be noted that factors are not created but are revealed based on correlations between surface attributes. Factors, which are unseen, can be intangible yet conceivable. For example, we could image factors within a given experiment being an individual's reading or writing ability when compared to one individual. These attributes aren't objective with respect to how we measure them, but when assessing a standardized test with a reading and writing section, for example, obviously affect a given person's score.

Limitations of Factor Analysis

Factor analysis can find a method of obtaining patterns in data generated even from random numbers. As such, one should keep in mind that if structure can be found in random data, than the patterns they appear to observe in their structured data could also be misconceived. Moreover, the structure found in the data ultimately is a derivative of the variables/data set inputted into the factor analysis. Simply stated, there are not objective patterns in data sets that make themselves apparent, and ultimately restructuring of data sets/variables can cause significant divergence in the results yielded by a factor analysis. As such, how one interprets the results of a factor analysis ultimately is far more subjective than it may seem. That said, it is recommended that factor analysis be used alongside statistical methods and/or the data be structured such that it conforms to assumptions known to be true within the domain of the problem being handled.

Handling Categorical Data

Among all of the difficulties that you might come across, one of the greatest challenges comes with handling and analyzing categorical data, or data that is numeric. Typically, we often encounter categorical data as a factor variable with different levels. This section talks about some common problems that will be encountered along with possible solutions, with considerations to keep in mind.

CHAPTER 8 ■ EXPERIMENTAL DESIGN AND HEURISTICS

Encoding Factor Levels

For example, let's say we have a data set where we are analyzing one variable, which is all of the streets in a given neighborhood. This is particularly interesting example because the streets could all be names, (such as "Maple Street," "Spruce Street," "Redwood Street," and so on), or they could all be numbers (1st Street, 2nd Street, 3rd Street, and so forth). If the streets are names, we can take the approach, to encode the streets by number. This is an easy way to give each variable a unique identifier, but it has limitations. Machine learning algorithms will interpret the levels as an indication of value rather than a unique identifier, which in essence gives no descriptive data about the "quality" of the observation. To be specific, if we label "Maple Street" as 1 and "Spruce Street" as 2, many algorithms might interpret Spruce Street to be of higher importance than Maple Street, when there is no evidence to determine this. When considering the case of the numbers, this same problem is present, but it's just implicit and not induced by label encoding. Another limitation of this technique is that if the encoded variable is highly correlated with other variables, multicollinearity might be introduced to the data set where it otherwise would not have existed.

Categorical Label Problems: Too Numerous Levels

In keeping with the example of using street names, we can imagine many cities where this would cause us to have a factor with hundreds or even thousands of individual streets. Although a variable with variation yields better results than a variable with absolutely none, this can also cause difficulties when performing model evaluation. As such, in these instances it can be a good idea to encode the variables and use a classification/regression tree or random forest model. Also, a suggested method is to encode the variables and use K-means clustering to get the cluster number, whereupon we replace the levels with this variable. Although this still in many ways has the bias of the encoded variable we discussed before baked into the clustering observation, it's nonetheless a method of reducing the levels effectively and should be explored when necessary.

Canonical Correlation Analysis (CCA)

Very closely related to PCA is canonical correlation analysis (CCA), a method of finding linear combinations of two variables such that they have the maximum possible covariance with each other. Typically, this is a data preprocessing technique and is appropriate in the same instances where multivariate linear regression would be used, but specifically when there are two sets of multivariable data sets that we want to examine the relationship between:

Given two vectors $X, Y \in \mathbb{R}^{m \times n}$ and directions $\alpha \in \mathbb{R}^m$ and $\beta \in \mathbb{R}^n$:

$$\alpha, \beta = \operatorname{argmax}\ cov(X\alpha, Y\beta)$$
$$\|X\alpha\|_2 = \|Y\beta\|_2 = 1$$

Wrappers, Filters, and Embedded (WFE) Algorithms

When assessing some of the more advanced variable selection techniques, we approach WFE algorithms. *Wrapper* algorithms are distinguished by running each feature subset possible over the data and evaluating the model performance, leading to the selection of a subset that performs the best with a given model. *Embedded* algorithms are explicitly written into the process of a model (L1 regularization with LASSO). *Filter* methods attempt to assess the merits of the feature by looking at the data itself rather than evaluating its performance on the methods alone.

Relief Algorithm

Designed by Aha, Kibler, and Albert in 1991, the *relief* algorithm is a feature-based weight algorithm inspired by instance-based learning. Each feature is assigned a weight denoting its relevance of the feature to the target. This algorithm is randomized and the updates of relevance values depend on the difference between the selected instance and the two nearest instances.

Algorithm

1) Given $\{(x_n, y_n)\}_{n=1}^{N}$, set $w^0 = \frac{1}{I}$, $T =$ number of iterations, $\sigma =$ kernel width, $\theta =$ stopping criterion.

2) For t = 1 : T
   a. Calculate pairwise distances w.r.t. w^{t-1}.
   b. Calculate P_m, P_h, and P_o.
   c. Update weights.
   d. If $\|w^t - w^{t-1}\| < \theta$, break.

Other Local Search Methods

Many of the algorithms addressed in the latter parts of this text will draw inspiration from, if not be directly related to, this subfield of optimization, typically used for computationally intensive optimization problems. We consider all possible solutions as being in a set we denote as the *feature space* or *search space*. The target is the global optimum that satisfies the optimization problem we seek to solve. Local search algorithms are initiated with a random element from the feature space and over each iteration chooses a new solution based on information yielded from the current neighborhood. After this stage, the algorithm will move to a given neighborhood in the nearest vicinity, but depending on the problem the search algorithm may choose more than one neighborhood.

Hill Climbing Search Methods

Prior to the development of machine learning that occurred in the 1980s and 1990s, *hill climbing* tended to be one of the more popular search methods. Hill climbing forms the motivation for many newer search methods described in this chapter and is still a useful technique with respect to parameter tuning. As with other search methods, hill climbing seeks to optimize an objective function within the locality of the current point. Hill climbing works best for functions that have one maximum or one minimum, so as to allow the algorithm to find the solution of the problem with relative ease. However, it faces many problems for functions with an abundance of local minima. To combat this, many different heuristics and methods, like random restarts to avoid local minima and stochastic neighborhood selection for the search trajectory, have been added to the basic hill climbing algorithm.

Genetic Algorithms (GAs)

Genetic algorithms are considered a direct outgrowth of the field of artificial intelligence, as they directly mock the process of evolution. In this algorithm, several subsets of the total feature space "evolve" so that the next subset is statistically better than the last iteration. The evolution process stops when a better subset can't be created, and the best of the subsets is chosen as the answer. The advantage of this algorithm over others is that genetic algorithms can accumulate information about a given feature space over many iterations, the process is inherently parallel so there is less probability of being stuck in local minima, and the algorithm in and of itself is relatively easy to understand. Among GAs' limitations are the fact that if there is an abundance of local optima, the GA doesn't always converge upon the global optimum. Also, this algorithm is likely not an optimal choice for deployment, because it has difficulty scaling, since the feature space size increases exponentially with the number of possible subsets.

Algorithm

- Choose an initial random population of solutions to choose from.
- Evaluate the solution based on some statistical criterion, such as MSE.
- Select the best individuals to be used.
- Generate new individuals by "mutating" the prior selected solutions.
- Evaluate the fitness of the new solutions.
- Stop when some criterion has been reached, such a loss tolerance.

Simulated Annealing (SA)

Among the heuristic techniques we will cover, one of the few probabilistic models assessed is SA. Inspired in name from *annealing* in metallurgy, SA imitates the effect of slowly cooling as slowly decreasing the probability of accepting worse solutions. We consider each solution as a state and that the neighborhood in which the algorithm can search progressively gets smaller. The algorithm converges upon a solution either after the feature space has been entirely searched, or another stopping criterion has been reached. 1

Algorithm

- T = Temperature = hot, Frozen = Stopping Criterion.
- While (Temperature != Frozen), move to a random point in the feature space and compute Δ Engery.
- If ΔEnergy < 0 or loss tolerance, accept new state with probability $e^{\frac{\Delta E}{T}}$ while system in thermal equilibrium at current T.
- If (E decreasing over last few iterations), $T = T(itertion + 1)$,
  Else T = Frozen.

The greatest difficulty with SA is the amount of parameter tuning required, which becomes time consuming as the amount of feature (and corresponding feature space) increase in size. Furthermore, there isn't a general baseline or rule of thumb for any of these parameters, further increasing the difficulty of this technique with heavily changing data sets. It should likely be considered more of a research technique than one you would deploy in an algorithm.

Ant Colony Optimization (ACO)

Ant colony algorithms (ACOs) are a set of optimization algorithms first introduced in the 1990s. Most useful for combinatorics problems, ACOs been used for tasks such as vehicle routing, computer vision, feature subset selection, quantitative finance, and other fields. The intuition is based on the activities of swarms of ants, and the ultimate goal is typically finding the best options given set of randomized options from a feature space. We can imagine an ant colony in this context to be a graph with nodes connected by edges, where each node represents one of the k features in the data set. The ant travels along the edges, "dropping pheromones" to attract more ants along subsequent iterations. The pheromones by design decay over time, but ants who travel along the shortest possible edges from point x to point y deposit more pheromones along a given path. Because ants are attracted to paths with more pheromones, this acts as the method by which an optimal solution is found. Each "ant" moves from one given state with a probability given by

$$p_{x,y}^k = \frac{\tau_{xy}^\alpha \left(\eta_{xy}^\beta\right)}{\sum_{j \in J_i^k} \tau_{xj}^\alpha \left(\eta_{xj}^\beta\right)}$$

where τ_{xy}^{α} = pheremone deposited on a given path, η_{xy}^{β} = the proportion of the distance from x : y to the sum of all paths' distances, β = tuning parameter, J_i^K = neighbor nodes that have not been visited

With pheromones updated as

$$\tau_{xy}(t+1) \to (1-\rho)\tau_{xy} + \sum_k \Delta\tau_{xy}^k$$

where τ_{xy} = pheremone deposited, and ρ = pheremone evaporation rate.

We denote $\Delta\tau_{xy}$ as the amount of pheromone dropped on a given path by an individual ant given by

$$\Delta\tau_{xy} = \begin{cases} Q/L_k, & \text{if kth ant travels along xy} \\ 0, & \text{elsewhere} \end{cases}$$

where Q = some constant, and L_k = a loss function defined by the user.

Although ACO problems are successful for instances in which there aren't very large numbers of features, and it typically performs better than simulated annealing and genetic algorithms, the problems become exponentially more difficult to solve with the addition of more nodes. In addition to this, although convergence is guaranteed, it is uncertain as to when convergence actually will occur.

Algorithm

- Initialize by creating full solution space.
- While stopping criterion not reached, position each ant at a given starting node.
- For each ant, choose next node via state transition rule.
- Apply pheromone update until every ant has reached a given solution.
- Evaluate each solution based on the selection criterion.
- Update best solution and apply pheromone update on this path.
- Repeat until convergence upon global optimum.

Variable Neighborhood Search (VNS)

VNS is a family of feature subset selection algorithms that are meant to deal with combinatorics challenges and henceforth provide guaranteed convergence. Developed in the late 1990s, VNS was inspired by the desire to find solutions for discrete and continuous optimization problems (linear and nonlinear programming problems are an example). The assumptions within VNS are that a local minimum with respect to a given neighborhood is in theory perhaps not the local minimum in another neighborhood, that local minima are relatively close to each other between one or more neighborhoods, and that a global minimum are local minima for all neighborhoods within the solution space. Among the algorithms available for VNS with respect to local search methods, there are related extensions that are more specified for given tasks. For feature-based selection, we will look at the filter-based algorithm for VNS.

Algorithm

- Find an initial solution S.
- Select the set of neighborhoods N_k for $k = 1, ..., j$ where $j = $ # of neighborhoods and a stopping criterion.
- Set k = 1 and generate a random point S' from the kth neighborhood of $S(S' \in N_k(S))$.
- Apply a search method such that the stopping criterion, if based on an objective function, is closer to being reached.
- If this solution is better than the prior solution, update the solution to the current one. Else, set k = k + 1 and retain the current solution.
- Continue until convergence upon global optimum or stopping criterion is reached.

Typically, we choose an information quotient or linear correlation as an evaluation function within these algorithms, but this is ultimately a parameter that can be altered. If you feel more advanced, feel free to implement your own deep learning and/or machine learning algorithms where instead of traditional gradient descent, you use one of the aforementioned search methods for parameter optimization. Although this can be difficult, it will provide you with an excellent exercise to get familiar with specific algorithms, while also helping you understand how performance is affected by specific operations within a given algorithm. That brings us to a similar topic with respect to refining existing machine learning algorithms: reactive search optimization.

Reactive Search Optimization (RSO)

RSO is a relatively new innovation in the field of optimization. It produces interesting implications that are worth mentioning for more advanced readers. The purpose of RSO lends itself to being of particular use to those who intend on creating machine learning platforms and tools that are intended for users who aren't as technically adept as the typical machine learning engineer. *Intelligent optimization* refers to a more specific area of research within RSO, but is nonetheless relevant. In this paradigm, we evaluate the effectiveness of different learning schemes. There are broadly three, which we will refer to as online, offline, and a combination of the two with varying proportions. This is the idea of implementing algorithms in different environments such that they have different search histories, which ultimately affect the action of the epoch that is currently in session.

Reactive Prohibitions

Prohibition-based techniques and *intelligent schemes*, in contrast to basic heuristics such as local search, are what provide the intellectual motivation for *tabu search*. Tabu search methods mainly gained their initial traction in the 1980s, and it has proved a large area of research given the fertile ground it occupies. Tabu search (TS) is particularly noteworthy when comparing it against local search methods because of the use of prior information gleaned from the data set, and how that influences the new iterations' outcomes. Assume that we have a feasible search space that is composed of binary strings with a length $L: \mathcal{X} = \{0,1\}^L$, X is the current configuration, and N(X) is the previous neighborhood. The following equation is related to tabu search that is prohibition-based

$$X^{t+1} = \text{BestNeighbor}\left(N_A\left(X^t\right)\right),$$

$$N_A\left(X^{t+1}\right) = \text{ALLOW}\left(N\left(X^{t+1}\right), X^0, \ldots, X^{t+1}\right)$$

where the ALLOW function selects a subset of $\left(N\left(X^{t+1}\right)\right)$ such that it is dependent on the entire search trajectory $X^0, \ldots, X^{t+1}$.

Tabu search algorithms are classified in many ways, but the initial distinguishing factor I'll elaborate on is deterministic versus stochastic systems within TS algorithms. The most basic form of tabu search is denoted as *strict tabu search*. In this algorithm, we observe N(X) to have the following value:

$$N_A\left(X^{t+1}\right) = \left\{X \in N\left(X^{t+1}\right) \text{ s.t. } X \notin \left\{X^0, \ldots, X^{t+1}\right\}\right\}$$

When introducing a prohibition parameter, T, that determines how long a move will remain prohibited after the execution of its inverse, we can obtain two algorithms that are different from strict tabu search. A neighbor is allowed if and only if it is obtained from the current point by applying a direction to the search such that its inverse has not been used during the last T iterations, such that

$$N_A\left(X^{t+1}\right) = \left\{X = \mu \circ X^t \text{ s.t. LastUsed}\left(\mu^{-1}\right) < (t-T)\right\},$$

where LastUsed() is the last usage time of move μ. If T changes with the iteration counter, the general dynamical system that generates the search trajectory comprises an additional evolution equation for T such that

$$T^t = \text{React}(T^{t-1}, X^0, \ldots, X^t,$$

CHAPTER 8 ■ EXPERIMENTAL DESIGN AND HEURISTICS

$$N_A(X^{t+1}) = \{X = \mu \circ X^t \; s.t. \, \text{LastUsed}(\mu^{-1}) < (t-T)\}, X^{t+1}$$
$$= \text{Best} - \text{Neighbor}(N_A(X^t))\}$$

For basic moves acting on binary strings, $\mu = \mu^{-1}$.

For stochastic models, we can substitute prohibition rules with probabilistic generation-acceptance rules with large probability for allowed moves, and small for prohibited ones. Stochasticity can increase the robust nature of TS algorithms. Stochasticity can limit or remove the benefit of memory-induced activity, as is the main draw to tabu search. *Robust tabu search* features a prohibition parameter that is randomly changed between an upper and lower bound during the search. In *fixed tabu search*, stochasticity can be added by randomly breaking ties, or the cost function decrease is obtained by more than one candidate of the Best-Neighbor() function. This same effect is observed when implementing stochasticity in reactive tabu search.

Fixed Tabu Search

Let us assume we have a search space X such that $X = [b_1, b_2 b_3]$ with a cost function $f([b_1,b_2,b_3]) = b_1 + 2b_2 + 3b_3 = 7b_1 b_2 b_3$, where b is a 3-bit string. The feasible points will be the edges of the 3-dimensional cube shown in Figure 8-8. The neighborhood of a given point is the set of points that are connected with edges. The point $X^0 = [0,0,0]$ with $f(X^0) = 0$ is a local minimizer due to the fact that other moves produce a higher cost.

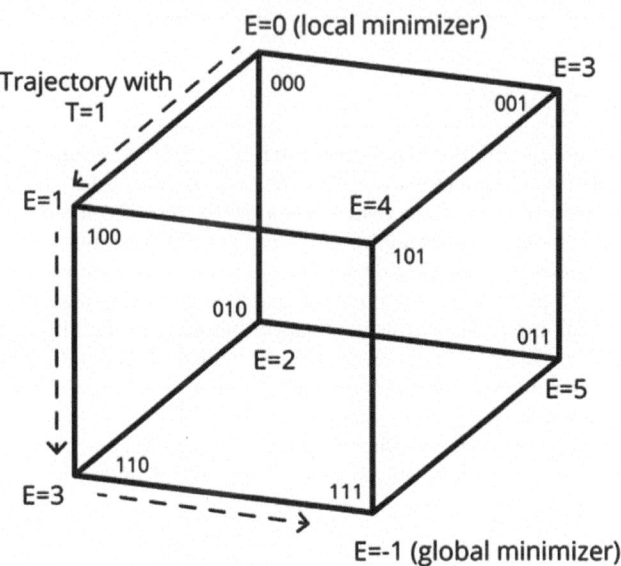

Figure 8-8. *A feature space with error function, E, and f value = [x,y,z], using tabu search*

We will define two parameters that will be of use to testing the efficiency of a given tabu search epoch, denoted as the Hamming distance and the minimum repetition interval. The *Hamming distance* describes the distance between the starting point and the most successful point along the search trajectory, and the *minimum repetition interval* describes the amount of times a similar move was visited along a given search trajectory. These parameters' equations are given by the following:

$$H(X^{t+1}, X^t) = \tau, \quad \tau \leq T+1,$$

$$X^{t+R} = X^t \Rightarrow R \geq 2(T+1)$$

Moving forward, we should direct our attention to avoiding attractors of the search trajectory, where we define *attractors* as local minima generated by deterministic local search. If the cost function is lower bounded, and starts from an arbitrary point, it will terminate at local minimizer. We also define what is known as an attraction basin. An *attraction basin* is composed of all points such that a deterministic local search trajectory starting from them terminates at a specific local minimizer. Deterministic search trajectories often suffer from being biased towards attraction basins and as such can yield a result that is not a global minimizer. To solve this, a given search point is kept close to a local minimizer that was found in the beginning of the search trajectory. After this, the search trajectory can search for better attraction basins with respect to reducing the cost function. As always, there are limitations that we must be conscious of. With tabu search, the difficulties that are most frequently encountered are the determination of an appropriate prohibition parameter and making the technique robust enough that it doesn't require tedious amounts of tuning from one context to another. This brings us to reactive tabu search, which has been proposed as a method of solving these problems.

Reactive Tabu Search (RTS)

Reactive tabu search (RTS) features a prohibition parameter that is determined through reactive mechanisms within the search trajectory. We initialize it with a value of 1 in the very beginning, but we add a deterministic aspect to how it changes. If there is evidence that diversification in the search trajectory is needed, T increases. Once this evidence isn't apparent, T decreases. Sufficient evidence for diversification in the search path is reached when we repetitively visit previous points along the search trajectory, as they are stored in the "memory" of the algorithm. Also, to avoid instances in which the algorithm is very rigidly stuck in an attraction basin, RTS has an escape mechanism. This is initiated when too many search trajectory configurations have been repeated in a given period and features a stochastic reconfiguration of the current search path.

The objective function, f, ultimately is where the information for the direction of the search trajectory comes from. As such, the following algorithms directly fall under this paradigm.

WalkSAT Algorithm

The WalkSAT algorithm can be understood as a more generalized version of the GSAT algorithm, which is a type of local search algorithm. In the algorithm, there are a set number of opportunities allowed for a given number of iterations to find a solution. During a given iteration, the algorithm chooses a variable between two criteria. After this point, the variable is put into the FLIP function where $\text{FLIP}(x_i) = (1 - x_i)$. The WalkSAT gets its power from doing less calculation than GSAT because it is considering fewer parameters at a given time. In addition to this, by a product of the clauses which determine variable picking, it thereby has the opportunity to solve a problem variable that could be preventing convergence upon the global optimum. Clause-weighting can also be incorporated into the WalkSAT algorithm, which gives new possibilities for parameter tuning and feedback loops produced upon, The following algorithm suggests weights as a method of encouraging more priority on solving the more difficult clauses. Difficult clauses are considered such after several configurations.

K-Nearest Neighbors (KNN)

KNN is considered to be instance-based learning, which features approximations of the function locally and all calculations happening after classification. It can also be used for regression, but often is described as a search method. Its main draws are the fact that it is relatively easy-to-understand and effective for cases in which there are irregularities in the pattern of data. These models, in the case of classification, are considered memory-based where we define k neighboring points that we want to consider. We use a Euclidean norm on the standardized data to determine the distance between a given point and its k neighbors. This equation is given as

$$d(x,y) = \sum_{i=1}^{N} \sqrt{(x_i - y)^2}$$

where $I = 1, 2, \ldots, N$, N = the total number of observations, x_i = ith observation, and y = the specific point we want to classify.

As K increases, typically we notice that the definition between classes becomes less rigid, leading to generally more robust models. Insofar as it relates to feature selection, KNN can be used as a data preprocessing technique often used alongside other search techniques for more refined feature selection. An example is given from a 2007 paper by Tahir, Bouridane, and Kurugollu in which they create a hybrid algorithm using a variant of tabu search and KNNs. The algorithm performs feature weighting and selection, yielding more accurate classification results. The pipeline occurs such that the features are selected and weighted via tabu search and classified via KNN. If we don't perform feature selection with tabu search, or feature selection at all, more noise is incorporated into the decision-making process for the KNN algorithm. As the case generally is, performing feature selection here helps the algorithm make more precise choices when classifying each observation.

Summary

This chapter was a kind of meta-heuristic on the entirety of the granular details discussed up until this point. Foremost, experimental design, feature selection, and A/B testing will be crucial to any data scientist's profession. The ability to properly structure the experiments by which you conduct models, improve upon their performance by modifying the inputs, and then quantitatively validate the results of a model are crucial. Chapter 9 discussed hardware solutions for those who are interested in creating a build for personal or professional use.

CHAPTER 9

Hardware and Software Suggestions

To apply the techniques explored in this book in a professional setting, hardware upgrades may become a consideration. In some cases, it might even be necessary to build a computer from the ground up. There are very few out-of-the-box ready builds, and the ones that do exist can cost a staggering amount of money. With that in mind, this chapter is intended to give readers a basic overview of the hardware components they should be most mindful of as well as provide general suggestions on hardware to purchase.

Processing Data with Standard Hardware

You may face many difficulties when operating on a relatively "vanilla" machine. When working on machine learning and deep learning problems with a large data set, it is generally recommended that you run most of your operations on subsets of the data and train in such a manner that the iterations times the size of the subset equals the size of the original data set. Although this merely provides an approximation of performance, it may be able to run your solution without crashing the interpreter due to lack of RAM.

It is also highly suggested that individuals with sufficient funds use Amazon Web Services (AWS). Professionally, Amazon is the go-to solution for cloud services and may even allow you to pick up a valuable skill set that many employers are eager to have. In short, you can pay to run instances of all the hardware you need in a cloud environment. Although for deployment purposes doing so can be extremely costly, for proof of concept or research using a cloud service like Amazon AWS can be a cost-efficient and easy solution to solving your problems for deep learning. If you need to implementing solutions as part of deploying an algorithm for a business or service, however, read on—the advice given in this chapter is a good starting point.

Solid State Drives and Hard Drive Disks (HDD)

A *hard drive disk* (HDD) is a storage device used to retain information even while the machine is not online. The main characteristics of an HDD are the amount of data it can store and the performance it provides. Since the mid 2000s, as I mentioned early in the book, the price of storage has dropped substantially, promoting a resurgence in interest in the science of machine learning and deep learning. This development makes it possible to store and collect substantial amounts of training data and/or trained models that you can update later moments or use for related tasks. Users should become familiar with the cases they want to tackle most often.

Graphics Processing Unit (GPU)

GPUs are one of the most frequently referenced pieces of hardware with respect to distinguishing machines that can deliver high-performance deep learning from machines that aren't specialized for deep learning). For deep learning, GPUs accelerate the processing of computations and are an integral part of the deep learning build. When compared with Central Processing Unit (CPU) computations, GPUs easily outperform CPUs and are where the bulk of computation occurs. You can build a unit with multiple GPUs, but be aware of the challenge of efficiently utilizing computing power when you do this. If you're not familiar with parallel computing, achieving such a build can be time-consuming to learn and implement correctly and invites spending an unknown amount of time, not to mention the time involved in designing and debugging software before algorithms/solutions can be effectively deployed.

There are packages in different languages to parallelize your code and improve performance. In R, I suggest you consider the parallel package, especially for performing the same task on a large amount of data. Rather than inputting the whole data set into an algorithm, it can be broken up such that the same task is performed in parallel with chunks of the data set, thereby making it more efficient. Where applicable, you should also implement the `lapply` function. This function takes a parameter and feeds it into a function, making performing complex operations much more computationally efficient than using nested loops.

My recommendations for GPUs (as of early 2017) focus on the following Nvidia models:

- Titan X
- GTX 680
- GTX 980

As of early 2017, Nvidia is one of the few companies devoting attention to developing GPUs specifically for the purpose of deep learning. (Note that AMD is partnering with Google to create deep learning hardware to be released sometime in 2017.) While this is likely not to be cost-effective for the average practitioner, for those in a professional context or with sufficient budgets, I suggest you review the specifications and performance reviews for AMD's FirePro S9300 x2 GPU when it releases.

Choosing a GPU depends on the type of problem you want to solve and how much memory you expect to consume in the process. Those using CNNs should expect to consume a great deal of memory, particularly in the process of training a given model. The physical storage for images and other data with deep learning is another consideration to keep in mind. Though both solid storage and virtual storage have dropped in price dramatically, you should set aside time to properly estimate the storage necessary.

Central Processing Unit (CPU)

The CPU instructs the computer on what operations should happen and where these operations should happen, in addition to performing very basic arithmetic, logical, and input/output functions. The CPU also works closely with the GPU to initiate function calls and initiate transfers of computations to the GPU. For deep learning–specific work, the number of CPU cores as well as CPU cache size are important. Most deep learning libraries rely on using a single CPU thread, and you can often perform just fine with one thread per GPU. However, using more threads per GPU will likely lead to better performance—take this fact in context with the task you intend to perform. For image-classification tasks, such the classic MNIST digit-classification task, I have found that using g2.2xlarge instances from AWS is more than sufficient, if I have difficulty using my local machine—it provides 1 GPU with 15 GB of RAM and 60 GB of SSD storage.

With respect to CPU cache size, there are several cache types with varying speeds. L1 and L2 tend to be quick, and L3 and L4 are slow. The purpose of the CPU cache is to help speed up computation via matching a key pair value. Most data sets encountered in a practical context are too large to fit into a CPU cache, so new data will be read in from the RAM on a given computer for each mini-batch. In the case of deep learning, most of the computation takes place in the GPU, so you needn't worry about buying CPUs that can handle this load. However, due to CPU cache misses, you may often see that the machine underperforms and you have latency issues. That leads to the core consideration with cache misses: RAM and the need for more of it so often in machine learning and deep learning.

Random Access Memory (RAM)

RAM stores frequently used program instructions such that the speed of programs increases because it stores data that will be read or written irrespective of its location within the RAM. As for the size of RAM you need, it should be comparable to the size of the GPU you're using. Using less RAM than the size of the GPU is likely to lead to latency issues that can cause problems particularly when training different networks such as CNNs. Using more RAM rather than less allows you to perform preprocessing and feature engineering much more easily than otherwise. It's easy to say, "Buy as much RAM as possible," but of course that's not always possible. However, you should consider investing a significant portion of available capital in this aspect.

Motherboard

The motherboard is the main circuit board, found in a variety of products besides personal computers. Its primary purpose is to facilitate communication between various components within a computer, and it holds the connectors between these components. Make sure the motherboard has enough PCIe ports to support the number of GPUs that will be installed in a given computer, as well as support all the other hardware components being chosen.

Power Supply Unit (PSU)

Power supply units convert alternating current electricity to regulated direct current power so that it can be used by the components within the computer. With regard to PSUs used for deep learning, be mindful to buy one that can service the number of GPUs you use if you use more than one. Deep learning can often require intensive periods of training, and the costs of running these instances should be minimized. The required watts for a given deep learning machine can be approximated by summing the watts of the GPU and CPUs while adding roughly 200 watts for the other components within the computer and variances in power consumption.

Optimizing Machine Learning Software

The major purpose of this chapter is to allow the reader to find where to focus their attention with respect to improving their machine. The end goal is to improve the performance of the software being tested and deployed, but part of that involves optimizing the software directly. To that end, before all other steps, I advise you to try to improve the algorithm you're using or find a better one when implementing a given solution. Optimal choice of algorithm and finding the most optimal implementation of said algorithm can be quite time-consuming. It might involve reading through a considerable amount of documentation, looking through the code for various functions in depth, and possibly doing experimentation. Although this book is intended for those who are relatively experienced in R and who are new to deep learning/machine learning, after reading through this text you should feel confident enough to begin creating your own implementations of various machine learning algorithms. Although time-intensive, doing so can teach you a great deal about the efficiency of different algorithms and their implementations.

A common debate currently revolves around which language to use. R is a very accessible language and great for proof of concept, particularly because its syntax allows for code to be written and tested quickly. Yet it can often prove cumbersome when trying to deploy the algorithms for anything that requires real-time applications—and particularly when trying to embed the software into other applications. If you intend on working in a professional context, keep that in mind when devising final solutions for anything. Typically, those looking to write for speed often do so in C++. This book doesn't cover C++, of course, or any of the packages in C++ for that matter, but readers should explore the myriad of libraries available in C++ for machine learning and deep learning.

Summary

This chapter should give readers a basic understanding of some of the most common concerns they should have when making a dedicated build for machine learning—or when trying to modify their existing hardware to better service their deep learning needs.

Chapter 10 dives into practical examples more heavily using machine learning and deep learning solutions.

CHAPTER 10

Machine Learning Example Problems

In this chapter we'll start applying the techniques discussed so far to practical problems you may potentially face. The data sets provided will either be generated from random data or will be from `https://github.com/TawehBeysolowII/ AnIntroductionToDeepLearning`. Note that you can also consult that URL for all code and data sets provided in the examples given in prior chapters.

In this chapter we will be exclusively examining machine learning problems. Though I can't cover every possible field and problem type, the focus on the examples here will to be address common scenarios users are likely to encounter.

I encourage you to view these final example chapters as tutorials for how to go from a data set (raw or processed) to a solution. Although these examples are feasible solutions, the most important aspect is applying the experimental design, feature selection, and model evaluation methodologies we've already discussed to solve problems effectively.

Problem 1: Asset Price Prediction

Quantitative finance is a field that continues to incorporate data science and machine learning techniques into its methods, specifically in the process of automated trading and market research. Although quantitative finance in and of itself is a field with a rich diversity and its own techniques, there are many broad analytic and mathematical concepts we can apply. For this example, we will be using the quantmod package to download financial data, and I'll walk you through how to predict asset prices. I'll also briefly explain how to create a trading strategy—specifically, a statistical arbitrage strategy. As always, backtesting these results is highly recommended prior to anyone applying these techniques. The purpose of this chapter is to provide an academic understanding of machine learning—it's not intended as a tutorial in quantitative portfolio management!

Let's assume you're a quantitative analyst at an asset management firm and you're tasked with reasonably predicting the returns of an asset that is in the S&P 500. Your managing director believes that there are ten other stocks that would be helpful in modeling the performance of this particular asset and that you should likely somehow use these in your analysis. The director gives no prescriptions particularly on what to use, besides suggesting using a machine learning approach to solving this problem.

Let's begin by defining the problem.

Problem Type: Supervised Learning—Regression

Any problem in which we're trying to predict discrete or continuous values is known as a *regression* problem. Because we have the answers, and we're trying to compare our proposed answers against the actual answers, this is a supervised learning problem. Specifically, we'll be trying to predict the returns of one asset, y, based on the returns of other assets, x. Let's start building the pipeline to solve this problem.

Typically, using the Yahoo! or Google Finance API is recommended for these tasks. For those particularly focused on the application of machine learning to quantitative finance, note that Yahoo! Finance's data has survivorship bias built in—that is, any companies that are now defunct cannot have their data accessed anymore. So, companies that were delisted for any reason are no longer stored in the database. This creates a problem because all strategies using this data won't reflect the worst possible downside had someone, for example, traded securities such as Bear Sterns of Lehman Brothers during the financial crisis. However, databases that hold data of companies that went bankrupt or are no longer listed can be found (Norgate Data is one example).

We'll begin by loading data using the Google Finance API, but will do so using the quantmod package. This package is recommended for any work requiring access to stock data, such as getting daily, monthly, or quarterly prices for various financial instruments, in addition to getting data on financial statements from publically listed companies.

Let's start walking through the code:

```
#Clear the workspace (1)
rm(list = ls())

#Upload the necessary packages (2)
require(quantmod)
require(MASS)
require(LiblineaR)
require(rpart)
require(mlbench)
require(caret)
require(lmridge)
require(e1071)
require(Metrics)
require(h2o)
require(class)
#Please access github to see the rest of the required packages!
```

```
#Summary Statistics Function
#We will use this later to evaluate our model performance (3)
summaryStatistics <- function(array){
  Mean <- mean(array)
  Std <- sd(array)
  Min <- min(array)
  Max <- max(array)
  Range <- Max - Min
  output <- data.frame("Mean" = Mean, "Std Dev" = Std, "Min" = Min,
  "Max" = Max, "Range" = Range)
  return(output)
}
```

In the preceding code, as always when using R, it's important to clear the workspace (1) when working with a new experiment. Then we load the required packages (2). The next function defined gives summary statistics on the arrays that we're analyzing (3). In this example, we'll be looking exclusively at MSE. This is to provide a simple example of how to evaluate machine learning models.

There are two approaches I often take:

- Evaluate several models in default mode and then perform parameter tuning on the best model.

- Perform parameter tuning one parameter at a time and then evaluate the tuned models against one another.

Here, I'll be performing the latter, though to a less intensive degree for the purpose of simplicity and explanation.

Description of the Experiment

The general pipeline we will create to solve this problem can be described as follows:

Data Ingestion → Feature Selection → Model Training and Evaluation → Model Selection

Specifically, in this problem we will try to predict the returns of Ford, ticker F, based on the returns of stocks we suspect accurately describe these returns (a mix of market indices and other stocks). The selection of our stock portfolio could be a study in and of itself, but in this instance we chose stocks that are related to the auto market (macro indicators and those tied to the energy industry). The assumption here is that stocks that track the performance of Ford are likely to be companies within the same industry, in related industries that service the auto market in some way, or describe greater implications about the economy at large.

Be aware that beyond the mathematics necessary to properly understand how to create machine learning models, it's necessary to provide these models with useful data. If we were to use features that are completely irrelevant to the problem being solved, we would be very unlikely to receive any useful results as output from a fitted model. As such,

CHAPTER 10 ■ MACHINE LEARNING EXAMPLE PROBLEMS

these assumptions we made to create our data set will help yield the better results prior to any fine tuning we perform on our algorithms:

```
#Loading Data From Yahoo Finance (4)
stocks <- c("F", "SPY", "DJIA", "HAL", "MSFT", "SWN", "SJM", "SLG", "STJ")
stockData <- list()
for(i in stocks){
  stockData[[i]] <- getSymbols(i, src = 'google', auto.assign = FALSE, from = "2013-01-01", to = "2017-01-01")
}

#Creating Matrix of close prices
df <- matrix(nrow = nrow(stockData[[1]]), ncol = length(stockData))
for (i in 1:length(stockData)){
  df[,i] <- stockData[[i]][,4]
}
#Calculating Returns
return_df <- matrix(nrow = nrow(df), ncol = ncol(df))
for (j in 1:ncol(return_df)){
  for(i in 1:nrow(return_df) - 1){
    return_df[i,j] <- (df[i+1, j]/df[i,j]) - 1
  }
}
```

In the preceding code, we pull the data from Yahoo! Finance (4). Unless this data is saved after initial download, you should have an active Internet connection—otherwise this part of the code won't execute properly. When calculating the returns of a given stock, you can think of returns as a derivative, but a simpler formula for a return based price is the following:

$$\text{Adjusted CloseR}_{x_t} = \left(\frac{P_{x_{t+1}}}{P_{x_t}} \right) - 1$$

(A)

Where x = stock x, y = stock y, t = time period (1,2, ... n),

n = number of observations, and P_{x_t} = Price of Stock x in period t

For the purpose of this experiment, and likewise in many such cases in quantitative finance, we calculate returns based on adjusted close prices (equation A). We call these *adjusted close prices* based on their reflecting any changes in the underlying stock price over time due to dividends, stock splits, or other financial adjustments that have nothing to do with the performance of the stock or market conditions. Here, we will be looking at daily returns. The selection of the time frequency is entirely up to the user and depends on the strategy being assessed. Generally speaking, high-frequency trading occurs multiple times within a day, and low-frequency trading occurs in increments significantly longer than a day.

We organize the data such that each column represents the returns of a given stock and each row represents the return on a given day. Figure 10-1 shows the head of the data set.

CHAPTER 10 ■ MACHINE LEARNING EXAMPLE PROBLEMS

```
            F         SPY         DJIA         HAL        MSFT         SWN         SJM
[1,]  0.019696990 -0.002259354 -0.001579823  0.016802086 -0.013396143  0.003598291 -0.005353604
[2,]  0.008172312  0.004391676  0.003274470  0.009363810 -0.018715611  0.025993396  0.011437504
[3,] -0.010316804 -0.002732758 -0.003790035  0.000000000 -0.001869807 -0.024170122 -0.003436768
[4,] -0.005956784 -0.002877293 -0.004142202 -0.012551241 -0.005245454 -0.023276576 -0.008899811
[5,]  0.008988766  0.002542050  0.004626067  0.008013288  0.005649777 -0.019553926  0.005612301
[6,]  0.026725973  0.007949585  0.006027400  0.011513202 -0.008988833  0.000000000  0.002790496
```

Figure 10-1. *Head of stock return data set*

Stock returns often work well with machine learning algorithms because they are all scaled similarly and represent a measure that is relative to all the observations within a given stock, as well as the universe of stocks available for analysis.

Feature Selection

When handling time series data, we often encounter multicollinearity. Because of this, PCA is a fair method to use for feature selection. We do so because there are likely features that are unnecessary to evaluate, and therefore noise need not be valuated, in addition to the fact that linear correlations among variables are high. So, evaluating features by their variance contributed is reasonable. The following shows the code that performs PCA:

```
#Feature Selection
#Removing last row since it is an NA VALUE
return_df  <- return_df[-nrow(return_df), ]
#Making DataFrame with all values except label IE all columns except for
Ford since we are trying to predict this
#Determing Which Variables Are Unnecessary
pca_df  <- return_df[, -1]
pca  <- prcomp(scale(pca_df))
cor(return_df[, -1])
summary(pca)
```

When executing the preceding code, we receive the results shown in Figures 10-2 and 10-3.

```
           SPY       DJIA       HAL       MSFT        SWN        SJM        SLG        STJ
SPY  1.0000000 0.9682642 0.5700102 0.59815000 0.28589655 0.4897106 0.5973075 0.4858736
DJIA 0.9682642 1.0000000 0.5335998 0.57874507 0.26338657 0.4780761 0.5664838 0.4396684
HAL  0.5700102 0.5335998 1.0000000 0.29254112 0.36502847 0.2233835 0.2819550 0.2501970
MSFT 0.5981500 0.5787451 0.2925411 1.00000000 0.06478864 0.2733258 0.3380027 0.2227678
SWN  0.2858966 0.2633866 0.3650285 0.06478864 1.00000000 0.0834507 0.1815691 0.1082383
SJM  0.4897106 0.4780761 0.2233835 0.27332580 0.08345070 1.0000000 0.3131153 0.2069674
SLG  0.5973075 0.5664838 0.2819550 0.33800270 0.18156911 0.3131153 1.0000000 0.3146444
STJ  0.4858736 0.4396684 0.2501970 0.22276779 0.10823825 0.2069674 0.3146444 1.0000000
```

Figure 10-2. *Correlation matrix for entire data set*

CHAPTER 10 ■ MACHINE LEARNING EXAMPLE PROBLEMS

```
Importance of components:
                          PC1    PC2    PC3    PC4     PC5     PC6     PC7     PC8
Standard deviation     1.9561 1.0354 0.8969 0.86337 0.80754 0.73743 0.57288 0.16610
Proportion of Variance 0.4783 0.1340 0.1006 0.09318 0.08151 0.06798 0.04102 0.00345
Cumulative Proportion  0.4783 0.6123 0.7129 0.80604 0.88755 0.95553 0.99655 1.00000
```

Figure 10-3. *Summary of principal components analysis (PCA) on data set*

In row 2 of Figure 10-3, you can see the proportion of the variance each principal component contributes to the data set. It must be stated for clarity that *principal components do not represent the features within the data set*. With that being said, we can consider principal component 1 to be a combination of features 1 through 8, PC 2 to be a combination of features 2 through 8, and so on. The general rule of thumb is to consider as insignificant principal components that contribute 1% or less to the total variance. When translating this to the data set, we would remove feature 8 within the data set. This same pattern of analysis should be extrapolated, but only when linear correlations between features are observed. Back in Figure 10-1, you can see generally moderate to strong linear correlations among the features, indicating that PCA is indeed an appropriate choice for features.

Model Evaluation

Now that we've preprocessed the data, let's consider our choices for algorithms. In this example, we'll evaluate a couple of different choices and evaluate the MSE on all of them. The number of models to choose is entirely up to you, but for this practical example I'll choose two. Furthermore, should you choose to evaluate statistics other than MSE, such as R Squared, it is reasonable to evaluate these measures relative to the goal of the experiment. That said, MSE should be and is the primary objective to minimize in regression models, and that should be the primary concern above all other evaluation methods.

Ridge Regression

Let's choose the first model: ridge regression. Here, we'll evaluate the MSE with respect to the value of the tuning parameter. In the following code, we begin by randomly sampling values from a normal distribution (5). These values will be used to pick the size of the tuning parameter, which we represent with K. The intuition behind this is that we'll sort the values from lowest to greatest and then compare the performance of our ridge regression model's MSEs by visualizing the error as we increase the tuning parameter:

```
#Ridge Regression
k <- sort(rnorm(100))(5)
```

CHAPTER 10 ■ MACHINE LEARNING EXAMPLE PROBLEMS

In the following code, we begin by cross validating our results so that we are evaluating generalities of model performance rather than testing our algorithm on the exact same data set (6). We choose to use a training and test set of equal size, by splitting the data in half:

```
mse_ridge <- c()
for (j in 1:length(k)){ (6)
    valid_rows <- sample(1:(nrow(return_df)/2))
    valid_set <- new_returns[valid_rows, -1]
    valid_y <- new_returns[valid_rows, 1]
#Ridge Regression (7)
    ridgeReg <- lmridge(valid_y ~ valid_set[,1] + valid_set[,2] +
    valid_set[,3] + valid_set[,4]
                                + valid_set[,5] + valid_set[,6], data =
as.data.frame(valid_set), type = type,  K = k[j])
    mse_ridge <- append(rstats1.lmridge(ridgeReg)$mse, mse_ridge)
}
```

We then move to fitting the data to the ridge regression model using the lmridge() function and then append the MSE to a vector entitled mse_ridge (7).

When executing the following code, we see the result shown in Figure 10-4:

```
#Plots of MSE and R2 as Tuning Parameter Grows
plot(k, mse_ridge, main = "MSE over Tuning Parameter Size", xlab = "K",
ylab = "MSE", type = "l",
    col = "cadetblue")
```

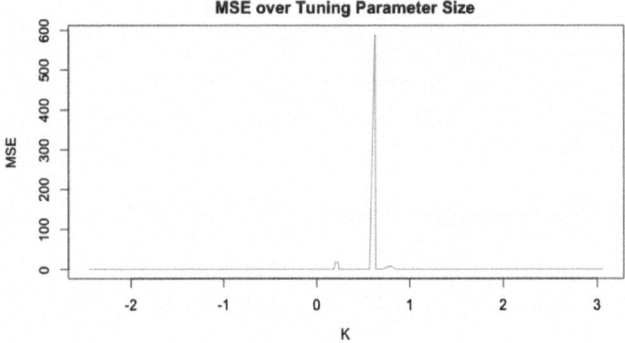

Figure 10-4. MSE over tuning parameter size

When looking at the plot, we see that the model performs best when our tuning parameter K is closest to the upper and lower bounds of the range displayed. Specifically, we'll choose to create a fitted model with a tuning parameter value of 1, as this K value yields a low MSE. When evaluating models it's important—in interviews, experiments, and for personal evaluation—to use plots to see the performance of the model with

CHAPTER 10 ■ MACHINE LEARNING EXAMPLE PROBLEMS

respect to some parameter value changing. This is useful for you as well as for other people who are using/evaluating your code. It will help to guide people through your thought process, and plots tend to be more engaging than looking at numerical outputs of code from a terminal.

Before we test our fitted model on data outside our validation set, let's show how we would tune another algorithm: the support vector regression (SVR).

Support Vector Regression (SVR)

The main parameter to tune here is the kernel function, which determines the shape of the hyperplane and therefore the shape of the regression line. When we execute the following code, we get the plot shown in Figure 10-5:

```
#Kernel Selection
svr_mse <- c()
k <- c("linear", "polynomial", "sigmoid")
for (i in 1:length(k)){
  valid_rows <- sample(1:(nrow(return_df)/2))
  valid_set <- new_returns[valid_rows, -1]
  valid_y <- new_returns[valid_rows, 1]
  
  SVR <- svm(valid_y ~ valid_set[,1] + valid_set[,2] + valid_set[,3] + valid_set[,4]
             + valid_set[,5] + valid_set[,6], kernel = k[i])
  svr_y <- predict(SVR, data = valid_set)
  svr_mse <- append(mse(valid_y, svr_y), svr_mse)
}

#Plots of MSE and R2 as Tuning Parameter Grows
plot(svr_mse, main = "MSE over Tuning Parameter Size", xlab = "K",
ylab = "MSE", type = "l",
     col = "cadetblue")
```

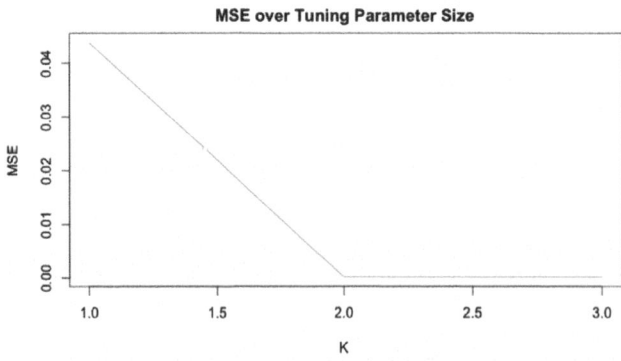

Figure 10-5. *SVR MSE with respect to kernel selection*

CHAPTER 10 ■ MACHINE LEARNING EXAMPLE PROBLEMS

When evaluating the output, we notice that following MSE values in Figure 10-5. The polynomial kernel yields the smallest MSE and therefore is our choice. Now, that we've trained both models, we'll predict out of sample using our tuned models. In a practical setting, you should likely fit more than two models and evaluate the performance. Because this process is exhaustive, I've condensed this example to comparing two models for the sake of explanation. Regardless, let's see the performance of our tuned models:

```
#Predicting out of Sample with Tuned Models
#Tuned Ridge Regression
ridgeReg <- lmridge(valid_y ~ valid_set[,1] + valid_set[,2] + valid_set[,3]
+ valid_set[,4]
                    + valid_set[,5] + valid_set[,6], data = as.data.
frame(valid_set), type = type,   K = 1)

y_h <- predict(ridgeReg, as.data.frame(new_returns[-valid_rows, -1]))
mse_ridge <- mse(new_returns[-valid_rows, 1], y_h)

#Tuned Support Vector Regression
svr <-   SVR <- svm(valid_y ~ valid_set[,1] + valid_set[,2] + valid_set[,3]
+ valid_set[,4]
                    + valid_set[,5] + valid_set[,6], kernel = "polynomial")
svr_y <- predict(svr, data = new_returns[-valid_rows, -1])
svr_mse <- mse(new_returns[-valid_rows, 1], svr_y)

#Tail of Predicted Value DataFrames
svr_pred <- cbind(new_returns[-valid_rows, 1], svr_y)
colnames(svr_pred) <- c("Actual", "Predicted")
tail(svr_pred)
ridge_pred <- cbind(new_returns[-valid_rows, 1], y_h)
colnames(ridge_pred) <- c("Actual", "Predicted")
tail(ridge_pred)
```

The preceding code uses the regression models we trained, except we set the parameter values based on which values yielded the lowest MSE. Although we fit the model to the training data, we're predicting on the test data. This is denoted by the fact that we're indexing from the return data frame using all the observations that we did not train the model against. When predicting on the test data set, the Figures 10-6 and 10-7 show the actual versus predicted stock values for each algorithm.

```
          Actual        Predicted
[499,] -0.018987372  -0.0123039729
[500,]  0.004838697   0.0041947405
[501,] -0.005617921  -0.0008327449
[502,] -0.011299488  -0.0120237661
[503,] -0.001632649  -0.0046792152
[504,] -0.008176593   0.0012242246
```

Figure 10-6. Tail of actual versus predicted data frame (SVR)

```
          Actual        Predicted
[499,] -0.018987372  -0.0121145917
[500,]  0.004838697   0.0010842864
[501,] -0.005617921  -0.0002152537
[502,] -0.011299488  -0.0096797502
[503,] -0.001632649  -0.0020402024
[504,] -0.008176593   0.0016012160
```

Figure 10-7. Tail of actual versus predicted data frame (ridge regression)

When evaluating the MSE of these algorithms, we receive the following results:

MSE for Support Vector Regression: 0.0002967161

MSE for Ridge Regression: 0.0002632815

Based on these results, it's reasonable to say that we should choose the ridge regression over the SVR based on the better MSE. You should feel free to work through the example given and use different feature selection algorithms, in addition to different algorithms altogether, when evaluating a solution. The purpose of this section, again, is to provide insight into how I generally approach these problems so that you may begin to develop your own methodology. Although there are general guidelines to model selection and tuning, everyone is free to perform this in their own way.

Let's now view a classification problem.

Problem 2: Speed Dating

In *speed dating*, participants meet many people, each for a few minutes, and then decide who they would like to see again. The data set we will be working with contains information on speed dating experiments conducted on graduate and professional students. Each person in the experiment met with 10–20 randomly selected people of the opposite sex (there were only heterosexual pairings) for four minutes each. After each speed date, each participant filled out a questionnaire about the other person. Our goal is to build a model to predict which pairs of daters want to meet each other again (that is, have a second date).

Problem Type: Classification

Any problem in which we're trying to determine binary or finite multinomial outcomes can be thought of as a classification problem. In this case, this will be a supervised problem, because we know the labels of the data beforehand, but we need to calculate them via a deterministic rule specific to this data set. A second date is only planned if both people in a given matching decide they would like to see the other person again. So, we'll create this column in the preprocessing stage of the data set:

```
#Upload Necessary Packages
require(ggplot2)
require(lattice)
require(nnet)
require(pROC)
require(ROCR)

#Clear the workspace
rm(list = ls())

#Upload the necessary data
data  <- read.csv("/Users/tawehbeysolow/Desktop/projectportfolio/
SpeedDating.csv", header = TRUE, stringsAsFactors = TRUE)

#Creating response label
second_date  <- matrix(nrow = nrow(data), ncol = 1)

for (i in 1:nrow(data)){
  if (data[i,1] + data[i,2] == 2){
    second_date[i]  <- 1
  } else {
    second_date[i]  <- 0
  }
}
```

As always, we begin the experiment by loading the necessary packages and clearing the workspace. Then we load the data and create a response label denoted second_date.

Now that we've gone through some initial preprocessing, let's describe and explore our data set. The features in this data set are as follows, from the first column through the last column:

- *Second_Date*: The response variable, y, for the data set which is binary. 1 = Yes (you would like to see the date again), 0 = No (you would not like to see the date again).

- *Decision*: The decision of the individual person, segregated by sex, as to whether they would like to go on a second date. 1 = Yes (you would like to see the date again), 0 = No (you would not like to see the date again).

- *Like*: Overall, how much do you like this person? (1 = not at all, 10 = like a lot).

- *PartnerYes*: How probable do you think it is that this person will say 'yes' for you? (1 = not probable, 10 = extremely probable).

- *Age*: Age.

- *Race*: Caucasian, Asian, Black, Latino, or Other.

- *Attractive*: Rate attractiveness of partner on a scale of 1–10 (1 = awful, 10 = great).

- *Sincere*: Rate sincerity of partner on a sale of 1–10 (1 = awful, 10 = great).

- *Fun*: Rate how fun partner is on a scale of 1–10 (1 = awful, 10 = great).

- *Ambitious*: Rate ambition of partner on a scale of 1–10 (1 = awful, 10 = great).

- *Shared Interest*: Rate the extent to which you share interests/hobbies with partner on a scale of 1–10 (1 = awful, 10 = great).

Preprocessing: Data Cleaning and Imputation

Note that in this data set there are NA observations. As mentioned, we have multiple tools to deal with this problem, but it's important for us to algorithmically find a way to handle this. We will tackle that prior to performing any feature transformation. The following code shows the process by which we handle NA data:

```
#Cleaning Data
#Finding NA Observations
lappend <- function (List, ...){
  List <- c(List, list(...))
  return(List)
}
na_index <- list()
for (i in 1:ncol(data)){
  na_index <- lappend(na_index, which(is.na(data[,i])))
}
```

First, we create a function that will let us append vectors to a list such that for each column, we have a vector of rows that indicate where the NA observations are. Given the nature of the data set, it's logical to impute the values using a method most reasonable given the data within that column/feature. Note that columns Second_Date, DecisionM, DecisionF, RaceM, and RaceF don't have any missing data. We're going to tackle the features that do have missing data.

We'll perform our data imputation using the expectation maximization (EM) algorithm described in Chapter 3. This is given in the amelia package, which can be installed from the R terminal. Before that, though, we must prepare our data slightly:

```
#Imputing NA Values where they are missing using EM Algorithm
#Step 1: Label Encoding Factor Variables to prepare for input to EM Algorithm
data$RaceM <- as.numeric(data$RaceM)
data$RaceF <- as.numeric(data$RaceF)

#Step 2: Inputting data to EM Algorithm
data <- amelia(x = data, m = 1, boot.type = "none")$imputations$imp1

#Proof of EM Imputation
na_index <- list()
for (i in 1:ncol(data)){
  na_index <- lappend(na_index, which(is.na(data[,i])))
}
na_index <- matrix(na_index, ncol = length(na_index), nrow = 1)
print(na_index)

 #Scaling Age Features using Gaussian Normalization
data$AgeM <- scale(data$AgeM)
data$AgeF <- scale(data$AgeF)
```

The EM algorithm can't handle *factors* (categorical variables). That means we must numerically encode these factors prior to their being inputted to the algorithm. After this, we execute the amelia function, which executes what we would like. Moving forward, we provide proof that there is no longer any NA data within this data set by indexing any NA values and then printing this output, yielding the result shown in Figure 10-8.

```
          [,1]      [,2]      [,3]      [,4]      [,5]      [,6]      [,7]      [,8]      [,9]
[1,] Integer,0 Integer,0 Integer,0 Integer,0 Integer,0 Integer,0 Integer,0 Integer,0 Integer,0
          [,10]     [,11]     [,12]     [,13]     [,14]     [,15]     [,16]     [,17]     [,18]
[1,] Integer,0 Integer,0 Integer,0 Integer,0 Integer,0 Integer,0 Integer,0 Integer,0 Integer,0
          [,19]     [,20]     [,21]     [,22]     [,23]
[1,] Integer,0 Integer,0 Integer,0 Integer,0 Integer,0
```

Figure 10-8. Displaying counts of NA values in cleaned data set

We've successfully removed all the NA observations and will perform the last bit of preprocessing before we move on to feature selection. Let's look at the distribution of ages with respect to both male and female. We code this as the following and receive the subsequent result:

```
#Scaling Age Features using Gaussian Normalization
summaryStatistics(data$AgeM)

Mean   Std.Dev Min Max Range
1 26.60727 3.509664  18  42    24
```

CHAPTER 10 ■ MACHINE LEARNING EXAMPLE PROBLEMS

```
summaryStatistics(data$AgeF)

  Mean    Std.Dev   Min  Max  Range
1 26.24317 3.977411  19   55    36

#Making Histograms of Data
hist(data$AgeM, main = "Distribution of Age in Males", xlab = "Age",
ylab = "Frequency", col = "darkorange3")
hist(data$AgeF, main = "Distribution of Age in Females", xlab = "Age",
ylab = "Frequency", col = "firebrick1")
data$AgeM <- scale(data$AgeM)
data$AgeF <- scale(data$AgeF)
```

When visualizing the distributions of the data using the hist() function, the code yields the results shown in Figures 10-9 and 10-10.

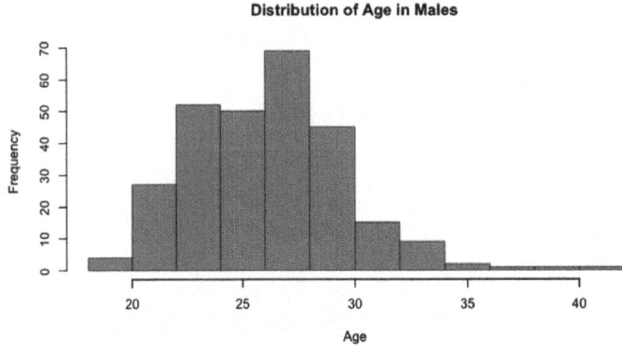

Figure 10-9. *Histogram of male ages*

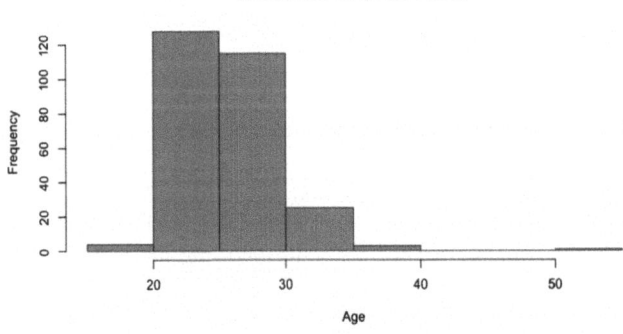

Figure 10-10. *Histogram of female ages*

The distributions of both female and male ages are positively skewed, meaning that the average is less than the median. However, note that there is significantly less variation in female ages in contrast to male ages. Although this also might serve as an insight we want to keep, you should glean the importance of displaying plots when exploring your data set and explaining what the information shows. This tends to be one of the most compelling ways to display information for people who aren't nearly as technical. For those who often find themselves making presentations, effective use of plots is a must. Finally, we end our data cleaning and preprocessing by performing Gaussian normalization on the age variables so that their inputs don't affect the accuracy of our classification models, because they are on different ranges than every other variable that isn't a numerical label.

Now that all the necessary preprocessing has been performed, we can approach the task of feature selection.

Feature Selection

This data set doesn't have an abnormally large number of observations, but 27 individual features likely makes for overkill and will unnecessarily weaken our machine learning algorithm's predictive power. As such, it is reasonable for us to eliminate unnecessary features, though we should be mindful of this process not necessarily being as straightforward as it appears.

When looking at the correlation matrix (the matrix is too large to be displayed here), we notice that there are generally weak to moderate linear correlations. We will likely be unable to get effective results from any models that rely heavily upon linear assumptions. When relating that to feature selection, we are similarly unlikely to get good results from using PCA. So, I chose to use a random forest to denote feature importance based on how much they affect the classification of an observation:

```
#Feature Selection
corr <- cor(data)

#Converting all Columns to Numeric prior to Input
for (i in 1:ncol(data)){
  data[,i] <- as.integer(data[,i])
}

#Random Forest Feature Selection Based on Importance of Classification
data$second_date <- as.factor(data$second_date)
featImport <- random.forest.importance(second_date ~., data = data,
importance.type = 1)
columns <- cutoff.k.percent(featImport, 0.4)
print(columns)
```

When executing the preceding code, the following columns are above the 0.4 threshold set for importance:

```
[1] "DecisionF"    "DecisionM"         "AttractiveM"    "FunF"     "LikeM"
[6] "LikeF"        "SharedInterestsF"  "AttractiveF"    "PartnerYesM"
```

These will be the features used in our training set, and we now can proceed to model training and evaluation.

Model Training and Evaluation

Now that we have a sufficiently reduced and transformed data set, it's time to go about the process of model selection. Because the function that determines the classification is not linear, we should look at functions that can handle this type of data. In the next problem, we'll use the following portfolio of algorithms:

- Logistic regression
- Bayesian classifier
- K-nearest neighbors

We'll tune each algorithm's parameters individually, evaluate the training set performance, and then predict out of sample. Once we've done this for all algorithms, we'll evaluate the results side by side and then choose the most optimal algorithm.

Method 1: Logistic Regression

It's suggested that when evaluating a portfolio classification algorithms you should always start with logistic regression. The reason is less because of the expectation for this to be the best algorithm, and more from the standpoint that this forms a baseline evaluation from which you can compare the different classification algorithms. In this experiment, we'll evaluate the performance of our models with respect to their AUC score, which is the area under the (ROC) curve:

```
#Method 1: Logistic Regression
lambda <- seq(.01, 1, .01)
AUC <- c()
for (i in 1:length(lambda)){
  rows <- sample(1:nrow(processedData), nrow(processedData)/2)
  logReg <- glm(as.factor(second_date[rows]) ~., data = processedData[rows, ],
    family = binomial(link = "logit"), method = "glm.fit")
  y_h <- ifelse(logReg$fitted.values >= lambda[i], 1, 0)
  AUC <- append(roc(y_h, as.numeric(second_date[-rows]))$auc, AUC)
}
```

CHAPTER 10 ■ MACHINE LEARNING EXAMPLE PROBLEMS

We start by altering the threshold that determines whether we classify an observation as a 1 or 0 based on the lambda parameter. We iterate over the algorithm and append the AUC score based on this parameter to the AUC vector. After this loop of iterations, we should evaluate the performance visually by using a plot. When plotting the AUC score vector over the lambda value, we write the following code and observe the output shown in Figure 10-11:

```
#Summary Statistics and Various Plots
plot(lambda[-1], AUC, main = "AUC over Lambda Value \n(Logistic
Regression)",
    xlab = "Lambda", ylab = "AUC", type = "l", col = "cadetblue")
```

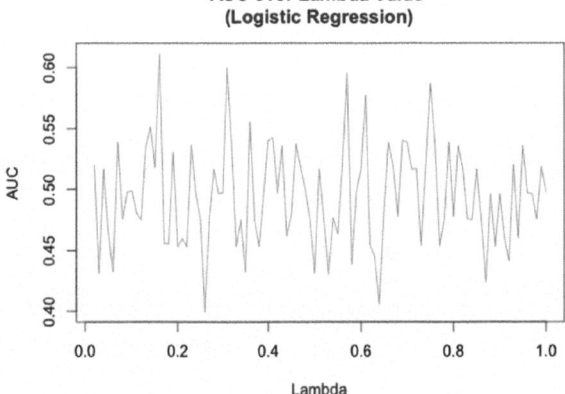

Figure 10-11. AUC over lambda value

We see that the AUC score is the highest when the lambda value is 0.15, so we'll use that lambda value. This is an example of how I would suggest you tune machine learning algorithms' parameters. Each parameter should be tuned individually so that you achieve a given objective, whether that is to minimize MSE or maximize AUC. In the logistic regression, the log odds threshold is really the only parameter we need to tune. We can view the performance of the tuned model over several iterations on the test set:

```
#Tuned Model
AUC <- c()
for (i in 1:length(lambda)){
  rows <- sample(1:nrow(processedData), nrow(processedData)/2)
  logReg <- glm(as.factor(second_date[rows]) ~., data = processedData[rows, ],
  family = binomial(link = "logit"), method = "glm.fit")
  y_h <- ifelse(logReg$fitted.values >= lambda[which(AUC == max(AUC))], 1, 0)
  AUC <- append(roc(y_h, as.numeric(second_date[-rows]))$auc, AUC)

}
```

CHAPTER 10 ■ MACHINE LEARNING EXAMPLE PROBLEMS

```
#Summary Statistics and Various Plots
plot(AUC, main = "AUC over 100 Iterations \n(Naive Bayes Classifier)",
     xlab = "Iterations", ylab = "AUC", type = "l", col = "cadetblue")
hist(AUC, main = "Histogram for AUC \n(Naive Bayes Classifier)",
     xlab = "AUC Value", ylab = "Frequency", col = "firebrick3")
```

We follow the same intuition as when tuning the machine learning algorithms, by collecting the AUC. The nature of the logistic regression is such that it fits a model upon each iteration rather than choosing the most optimal regression solution, as some algorithms do. When plotting the AUC vector with respect to the iterations over time and plotting a histogram of the AUC vector, we observe the results shown in Figures 10-12 and 10-13.

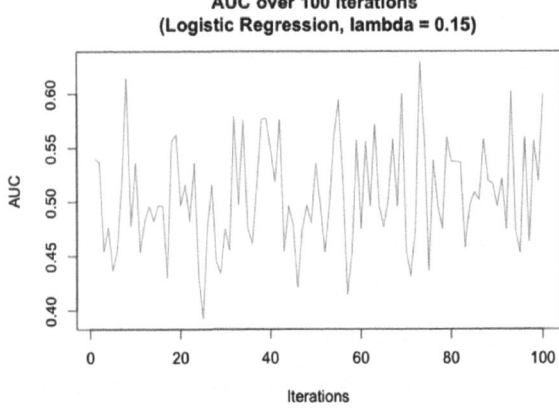

Figure 10-12. Logistic regression AUC over 100 iterations

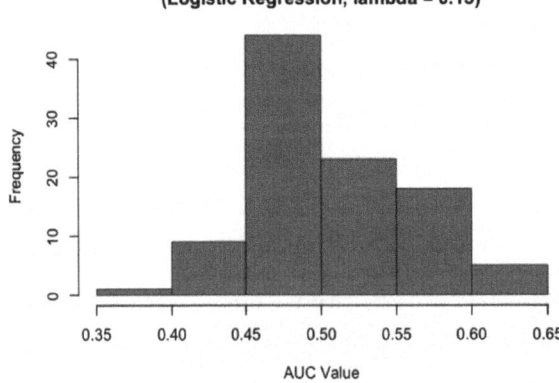

Figure 10-13. Logistic regression AUC histogram over 100 iterations

Numerically, we can summarize this vector using the following function:

```
summaryStatistics(AUC)
```

```
  Mean       Std.Dev     Min         Max         Range
1 0.5063276  0.04964798  0.3920711   0.6297832   0.2377121
```

We'll keep these values in mind moving forward. When analyzing them as is, logistic regression is an insufficient classifier. Typically, we would like to see AUC scores be at least .70, because a score of .50 indicates that the model is correct only 50% of the time. Less than .50 is not optimal and arguably means that we should consider this classifier insufficient.

Method 3: K-Nearest Neighbors (KNN)

This is a fairly simple classification algorithm described in detail in Chapter 3. The purpose in picking this algorithm relative to another probabilistic algorithm is to create a diverse algorithm portfolio such that we can infer which types of algorithms are best suited to this task. As a note to the reader, the K-NN algorithm in the class package yields the classifications from the test data. To train your algorithm on the training data only, use the same data that you assign to the "train" argument:

```
#Method 3: K-Nearest Neighbor
#Tuning K Parameter (Number of Neighbors)
K <- seq(1, 40, 1)
AUC <- c()
for (i in 1:length(K)){
  rows <- sample(1:nrow(processedData), nrow(processedData)/2)
  y_h <- knn(train = processedData[rows, ], test = processedData[rows,],
  cl = second_date[rows], k = K[i], use.all = TRUE)
  AUC <- append(roc(y_h, as.numeric(second_date[rows]))$auc, AUC)
}

#Summary Statistics and Various Plots
plot(AUC, main = "AUC over K Value \n(K Nearest Neighbor)",  xlab = "K", 
ylab = "AUC", type = "l", col = "cadetblue")
```

When looking at the plot of the AUC over K-value chart, we see the results shown in Figure 10-14.

CHAPTER 10 ■ MACHINE LEARNING EXAMPLE PROBLEMS

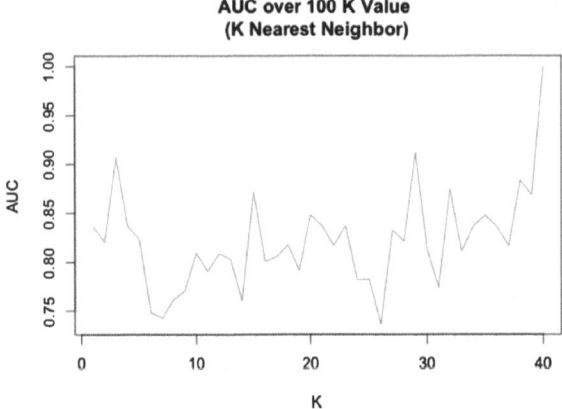

Figure 10-14. KNN classifier AUC over 100 iterations

The AUC score in the training phase is generally impressive for all the values, but it's reasonable to choose a lower K value than a large one to prevent overfitting. As such, we will choose a K of 3. Let's observe the AUC scores on the test set with our tuned model, as shown in Figures 10-15 and 10-16.

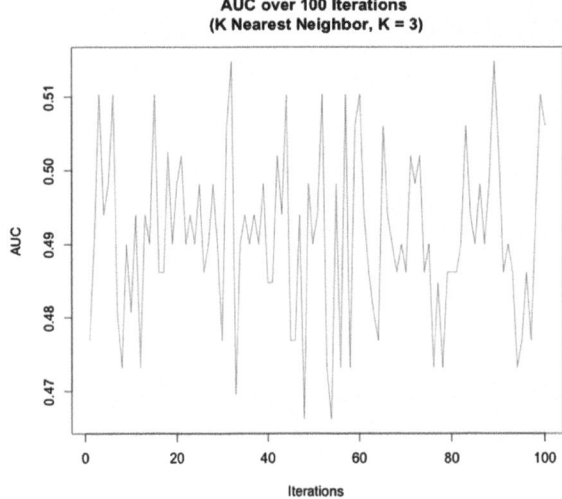

Figure 10-15. KNN AUC over 100 iterations on test set

190

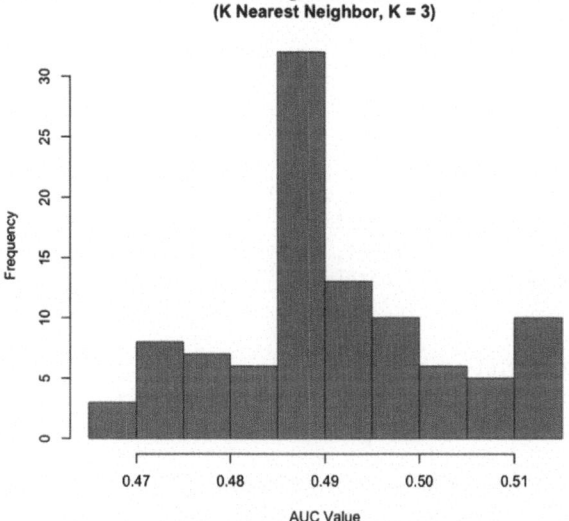

Figure 10-16. KNN AUC over 100 iterations on test set histogram

Numerically, we evaluate the AUC vector as the following:

summaryStatistics(AUC)

```
Mean        Std.Dev      Min         Max         Range
1 0.445006  0.01126862   0.4257075   0.4663915   0.04068396
```

Finally, we predict out of sample and observe the following results:

```
#Predicting out of Sample
y_h <- knn(train = processedData[rows, ], test = processedData[-rows, ],
cl = second_date[-rows])
roc(y_h, as.numeric(second_date[-rows]))$auc
```

Area under the curve: 0.4638

We see a stark drop-off from the training set to the test set, in addition to the test set performance being objectively poor.

Method 2: Bayesian Classifier

I suspect that occurrence of a second date can be modeled by Bayesian estimators, so the first model we'll begin with is the Bayesian classifier. In the following code, first we perform two-fold cross-validation on the data set so that we evaluate the performance on the training set. In this particular model, very little tuning needs to occur, so we'll just observe the performance of the model over 100 iterations:

```
#Method 1: Bayesian Classifier
AUC <- c()
for (i in 1:100){
  rows <- sample(1:nrow(processedData), 92)
  bayesClass <- naiveBayes(y = as.factor(second_date[rows]),
  x = processedData[rows, ], data = processedData)
  y_h <- predict(bayesClass, processedData[rows, ], type = c("class"))
  AUC <- append(roc(y_h, as.numeric(second_date[rows]))$auc, AUC)
}

#Summary Statistics and Various Plots
plot(AUC, main = "AUC over 100 Iterations \n(Naive Bayes Classifier)",
     xlab = "Iterations", ylab = "AUC", type = "l", col = "cadetblue")

hist(AUC, main = "Histogram for AUC \n(Naive Bayes Classifier)",
     xlab = "AUC Value", ylab = "Frequency", col = "cadetblue")

summaryStatistics(AUC)
```

When executing the code, we append the AUC score to the vector AUC, as shown in the preceding code that is looped over for 100 iterations. A line plot and histogram of this vector is shown in Figures 10-17 and 10-18.

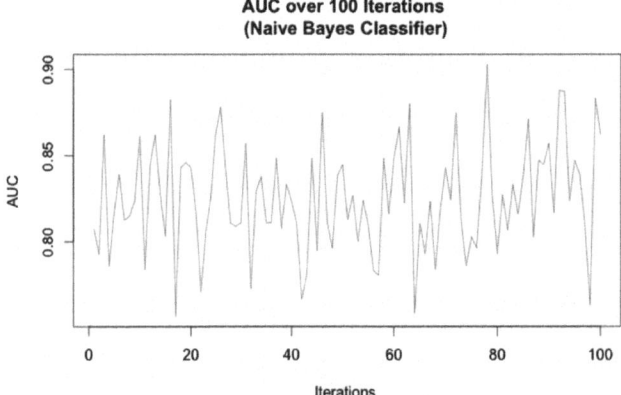

Figure 10-17. Bayes classifier AUC performance over 100 iterations

CHAPTER 10 ■ MACHINE LEARNING EXAMPLE PROBLEMS

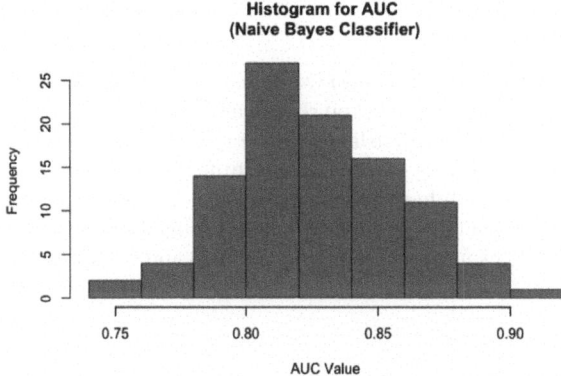

Figure 10-18. Bayes classifier AUC histogram over 100 iterations

We observe that the AUC scores have a slight right skew in their distribution and that the majority of the AUC scores are distributed within a relatively tight band of one another. When looking at the raw numerical data, we observe the following:

```
Mean         Std.Dev      Min          Max          Range
1 0.8251087  0.03142345   0.7567568    0.9027778    0.146021
```

These AUC scores yielded are more than generally acceptable for a model we choose, though we should still evaluate the performance of the model out of sample to be certain of how stable this process is:

```
#Predicting out of Sample
y_h <- predict(bayesClass, processedData[-rows, ], type = c("class"))
roc(y_h, as.numeric(second_date[-rows]))$auc
```

After executing the following code, we observe the following AUC score: area under the curve: 0.8219. This is acceptable within the distribution of the data yielded from the training set, with this AUC score trending towards the mean of the data.

When evaluating the solutions chosen, I strongly suggest choosing the Bayesian classifier given its stability from the training to the test set and superior AUC score above all other methods. In a practical setting, we would use the predictions out of the sample data to help influence our decision-making processes. In a professional context, this might include targeted marketing or recommendations to different users based on their dating profiles.

Summary

You now have a brief but comprehensive view into how I would recommend applying the concepts I've explained in the previous chapters. You should also note that although I've had success in implementing machine learning algorithms using this general process/methodology, this isn't the only way of training/tuning machine learning models. Nevertheless, I strongly emphasize the use of metrics and plotting the performance of the models with respect to these metrics when tuning different parameters. Chapter 11 will look at use examples of how to implement and use various deep learning models.

CHAPTER 11

Deep Learning and Other Example Problems

Now that I've sufficiently covered how to use and apply machine learning concepts, we should finally dive into applying and coding deep learning models using R. This can seem like a daunting task, but don't be intimidated. If you have been able to code everything successfully in this book, it's just a matter of adjusting to new packages. We will discuss a variety of deep learning examples, but will begin by dealing with simpler models and then eventually going on to more complex models. The purpose of these exercises is twofold:

- To show how to construct these models or access them from various packages
- To give examples of how they could be used in a practical concept

Autoencoders

Many of the other models described in the deep learning chapters of the book are relatively straightforward when it comes to how to use them, but I have found that the use of autoencoders does *not* become automatically clear. Therefore, I want to explore a use case in which the use of autoencoders is made abundantly clear in a practical context. Let's consider a case in which we would like to use an autoencoder to improve the performance of a classification algorithm from Chapter 10. Specifically, I mean the classification problem we walked through, in which we were trying to determine whether a pair of individuals will go on a second date or not based on several features. Let's begin by working with the Bayesian classifier:

```
#Bayes Classifier
#Bayes Classifier
AUC <- c()
for (i in 1:100){
  rows <- sample(1:nrow(processedData), 92)
  bayesClass <- naiveBayes(y = as.factor(second_date[rows]),
x = processedData[rows, ], data = processedData)
```

CHAPTER 11 ■ DEEP LEARNING AND OTHER EXAMPLE PROBLEMS

```
    y_h <- predict(bayesClass, processedData[rows, ], type = c("class"))
    AUC <- append(roc(y_h, as.numeric(second_date[rows]))$auc, AUC)
}

summaryStatistics(AUC)
curve <- roc(y_h, as.numeric(second_date[rows]))
plot(curve, main = "Bayesian Classifier ROC")
```

When executing the preceding code, it yields what is shown in Figure 11-1.

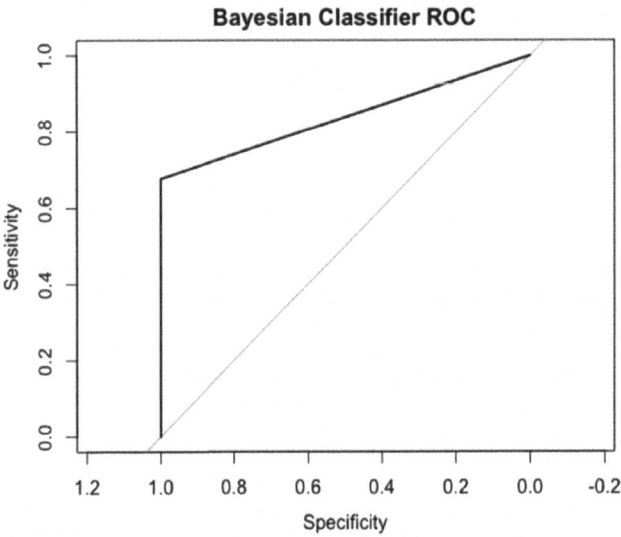

Figure 11-1. *ROC plot for Bayesian classifier*

We observe AUC scores of this model when collecting sample statistics:

```
Mean       Std.Dev      Min        Max      Range
0.8210827  0.02375922   0.7571429  0.875    0.1178571
```

These are objectively good scores. However, for the purpose of this example, we're going to use an autoencoder to help improve the performance of this model even further. This is where I introduce h2o. h2o produces a deep learning framework for R (along with other languages) that you will find useful for implementing many models. I encourage you to search through the documentation, because some implementations of deep learning models are hard to find (not to mention finding robust implementations). So let's initialize h2o and use an autoencoder:

```
#Autoencoder
h2o.init()
training_data <- as.h2o(processedData, destination_frame = "train_data")
```

CHAPTER 11 ■ DEEP LEARNING AND OTHER EXAMPLE PROBLEMS

```
autoencoder <- h2o.deeplearning(x = colnames(processedData),
 training_frame = training_data, autoencoder = TRUE, activation = "Tanh",
 hidden = c(6,5,6), epochs = 10)
autoencoder
```

h2o is similar to TensorFlow in that each session must be initialized. After this is initialized, whatever data passes through the models used must be transformed into an h2o-friendly format. We perform that transformation on our training data. Our autoencoder has three hidden layers, each of which has six, five, and six respective neurons within the given layers (denoted by the "hidden" argument within the h2o.deeplearning() function. We use tanh as our activation function. Upon executing the following code, we see what is shown in Figure 11-2.

```
H2OAutoEncoderModel: deeplearning
Model ID:  DeepLearning_model_R_1494853800072_2
Status of Neuron Layers: auto-encoder, gaussian distribution, Quadratic loss, 194 weights/biases, 7.0 KB, 2,760 training samples, mini-batch size 1
  layer units  type dropout       l1          l2 mean_rate  rate_rms momentum mean_weight weight_rms  mean_bias
1     1     9 Input  0.00 %
2     2     6  Tanh  0.00 % 0.000000 0.000000   0.018243  0.006420 0.000000    0.071839   0.385598   0.004520
3     3     5  Tanh  0.00 % 0.000000 0.000000   0.011672  0.003513 0.000000    0.039014   0.467069  -0.002555
4     4     6  Tanh  0.00 % 0.000000 0.000000   0.006541  0.002898 0.000000   -0.021580   0.422184  -0.001068
5     5     9  Tanh         0.000000 0.000000   0.008244  0.003908 0.000000   -0.006481   0.398464   0.010389
  bias_rms
1
2 0.017370
3 0.008162
4 0.018702
5 0.007152

H2OAutoEncoderMetrics: deeplearning
** Reported on training data. **

Training Set Metrics:
=====================

MSE: (Extract with `h2o.mse`) 0.01629403
RMSE: (Extract with `h2o.rmse`) 0.1276481
```

***Figure 11-2.** Summary of autoencoder function*

Note the MSE values. Because we're trying to recreate inputs of a function, this becomes a regression task. So we evaluate the effectiveness of this algorithm using the traditional regression statistics (MSE and RSME). Let's take a close look at the MSE yielded here and view the MSE with respect to the index of the data frame that holds the training data:

```
#Reconstruct Original Data Set
syntheticData <- h2o.anomaly(autoencoder, training_data, per_feature = FALSE)
errorRate <- as.data.frame(syntheticData)

#Plotting Error Rate of Feature Reconstruction
plot(sort(errorRate$Reconstruction.MSE), main = "Reconstruction Error Rate")
```

The h2o.anomaly() function uses the autoencoder to detect *anomalies*, which statistically we define as observations whose MSE during the reconstruction process are significantly higher than others. When executing the preceding code, we yield Figure 11-3.

CHAPTER 11 ■ DEEP LEARNING AND OTHER EXAMPLE PROBLEMS

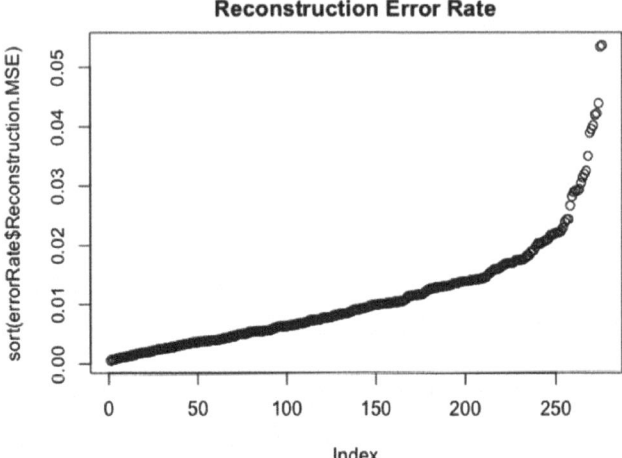

Figure 11-3. Plot of reconstruction error

We can see that there is a steady increase of the MSE but also a sharp increase from the index level 225 through the end of the training data. We can reasonably state that the outliers are generally these last inputs. With this in mind, we'll use a threshold determined by the MSE when segregating outliers from non-outliers into their respective subsets. We seek to train our Bayesian classifier by fitting our model to these subsets and seeing how the performance of the model, with respect to the AUC score, improves (or doesn't):

```
#Removing Anomolies from Data
train_data <- processedData[errorRate$Reconstruction.MSE < 0.01, ]

#Bayes Classifier
AUC <- c()
for (i in 1:100){
  rows <- sample(1:nrow(processedData), 92)
  bayesClass1 <- naiveBayes(y = as.factor(second_date[rows]), x =
  processedData[rows, ], data = processedData)
  y_h <- predict(bayesClass1, processedData[rows, ], type = c("class"))
  AUC <- append(roc(y_h, as.numeric(second_date[rows]))$auc, AUC)
}

#Summary Statistics
summaryStatistics(AUC)
```

We follow the same general steps we followed in Chapter 10 with respect to model training, collecting samples of the AUC statistic over 100 trials. The only difference here is that we're using a subset of the data with respect to the index values that fall below the MSE threshold. When looking at the summary statistics, we observe the following:

```
Mean       Std.Dev      Min     Max         Range
0.8274664  0.03076285   0.75    0.9117647   0.1617647
```

When comparing the distribution of our results to the original model, we observe a slightly higher mean, a higher max. However, we also observe a lower minimum. Therefore, the range and standard deviation of our results increase. Let's evaluate our results when we only look at anomalies:

```
########################################################################
#Using only Anomalies in Data Set
train_data <- processedData[errorRate$Reconstruction.MSE >= 0.01, ]

#Bayes Classifier
AUC <- c()
for (i in 1:100){
  rows <- sample(1:nrow(processedData), 92)
  bayesClass2 <- naiveBayes(y = as.factor(second_date[rows]),
  x = processedData[rows, ], data = processedData)
  y_h <- predict(bayesClass2, processedData[rows, ], type = c("class"))
  AUC <- append(roc(y_h, as.numeric(second_date[rows]))$auc, AUC)
}

#Summary Statistics
summaryStatistics(AUC)
```

When executing the preceding code, we see the following results:

```
Mean       Std.Dev      Min        Max         Range
0.8323727  0.03168166   0.7692308  0.9107143   0.1414835
```

Here we observe that this distribution contains the highest mean and minimum, with moderate results with respect to range and standard deviation. When choosing between the two data sets, I would argue for using the second subset in this instance due to the superior AUC score performance on average—and given the fact that at a minimum, we can still expect a higher score.

The importance of this technique lies in the fact that it is an effective method by which you can fit superior models on subsets of data. This will be extremely handy if you find you have a data set that is smaller than you would like. There are times when you can find yourself stuck trying to tweak a model whose performance is slightly unsatisfactory, despite using proper cross-validation techniques, data preprocessing techniques, and parameter tuning techniques. In instances where this is due to lack of data, this technique

CHAPTER 11 ■ DEEP LEARNING AND OTHER EXAMPLE PROBLEMS

would be the first I tried to use prior to trying to acquire more data. As for the final step in our experiment, let's use the fitted models and see how they perform out of sample:

```
#Fitted Models and Out of Sample Performance
AUC1 <- AUC2 <- c()

for (i in 1:100){
  rows <- sample(1:nrow(processedData), 92)
  y_h1 <- predict(bayesClass1, processedData[-rows,], type = c("class"))
  y_h2 <- predict(bayesClass2, processedData[-rows,], type = c("class"))
  AUC1 <- append(roc(y_h1, as.numeric(second_date[-rows]))$auc, AUC1)
  AUC2 <- append(roc(y_h2, as.numeric(second_date[-rows]))$auc, AUC2)
}
summaryStatistics(AUC1)
summaryStatistics(AUC2)
```

When executing the preceding code, we see the results for the model fitted against the subset without and with only anomalies respectively in Figures 11-4 and 11-5:

Mean	Std.Dev	Min	Max	Range
0.7890102	0.01468805	0.75	0.8194444	0.06944444
Mean	Std.Dev	Min	Max	Range
0.8303613	0.01506222	0.7957983	0.8688836	0.07308532

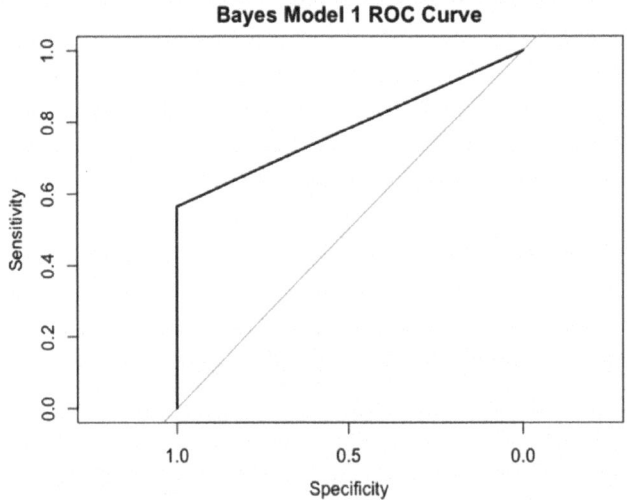

Figure 11-4. *ROC curve for Bayes model without anomalies (AUC : 0.7821)*

CHAPTER 11 ■ DEEP LEARNING AND OTHER EXAMPLE PROBLEMS

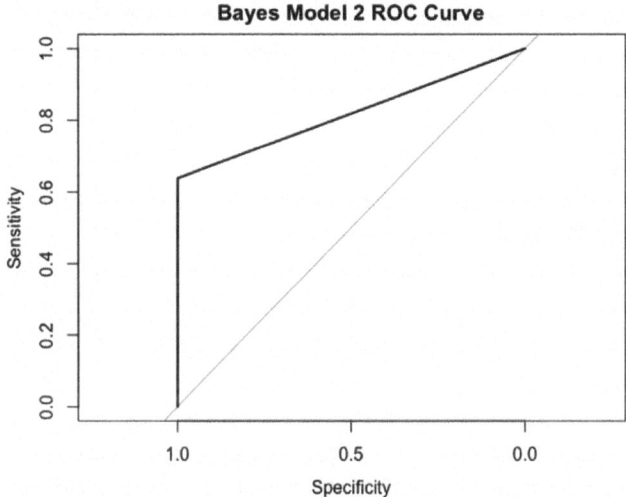

Figure 11-5. *ROC curve for Bayes model w/o anomalies (AUC: 0.8188)*

When reviewing the results from our experiment, it has become abundantly clear that the second model, fitted with only anomalies, produces a markedly better model than the model fit with observations that aren't anomalies. But before we become entirely convinced that we should use the second model, let's quickly perform a two-sided hypothesis test on using data from both of these models.

Being that we sampled our results 100 times, we can safely use a Z-test. As such, we set the Z-test parameters as shown in the following code:

```
#Two Sided Hypothesis Test
require(BSDA)

z.test(x = AUC1, y = AUC2, alternative = "two.sided", mu = mean(AUC2) - mean(AUC1),
            conf.level = 0.99, sigma.x = sd(AUC1), sigma.y = sd(AUC2))
```

When executing the preceding function, it yields the output shown in Figure 11-6.

```
	Two-sample z-Test

data:  AUC1 and AUC2
z = 1.7266, p-value = 0.08424
alternative hypothesis: true difference in means is not equal to -0.002053357
99 percent confidence interval:
 -0.004073352  0.008180066
sample estimates:
mean of x mean of y
0.8014105 0.7993571
```

Figure 11-6. *Two-sided hypothesis test results*

Statistically, within a 99% confidence interval, we have determined that the results of the two models are statistically different from one another and therefore we can confidently choose the second Bayesian model fitted, knowing that it is the superior model.

Convolutional Neural Networks

When I discussed CNNs in Chapter 5, I showed the power of this model by discussing the MNIST digit recognition use case. Although that was at one point the primary use case of CNNs, they are now currently being used for increasingly more difficult and complex tasks. Now I'd like to explore a use case in which we're trying to distinguish between different objects of significantly more complexity than handwritten digits. In this tutorial, we'll be using the Caltech 101 dataset, which contains 101 object categories with between 60 and 800 images in each category. We'll take various images from each category, doing so in such a way that we get diversity of images without picking starkly different pictures. We'll be choosing between images of guitars and laptops. Sample of theses photos are shown in Figures 11-7 and 11-8.

CHAPTER 11 ■ DEEP LEARNING AND OTHER EXAMPLE PROBLEMS

Figure 11-7. Photo of guitar

Figure 11-8. Photo of laptop

These images are pieces of technology, but they're distinctly different from each other in such a way that we would expect a human to be able to distinguish them. Let's now discuss how we should prepare our data for the CNNs.

Preprocessing

Working with image files requires a particular type of preprocessing that we haven't discussed in detail yet, mainly because image recognition and computer vision is a very specific subfield of computer science. It would be wise to seek other texts to build upon your understanding of computer vision, but this passage will give you a basic overview. We're working with color images, each with dimension x, y, z, where x and y are specific to each photo but z is always 3. Image files, insofar as a computer understands them, are three layers of matrices stacked on top of each other, with each pixel being an individual entry in that matrix. For this task, I recommend you use the EBImage package so you can grayscale and resize images. To help with the training time of the neural network, we'll be resizing images so they're smaller, and therefore the neural network takes in less data. But let's walk through our preprocessing step by step:

```
#Loading required packages
require(mxnet)
require(EBImage)
require(jpeg)
require(pROC)

#Downloading the strings of the image files in each directory
guitar_photos <- list.files("/file/path/to/image")
laptop_photos <- list.files("/file/path/to/image")
```

The Caltech library is organized into directories with multiple levels, so be mindful when trying to access these images in an automated fashion. All the directories for each category have the same format for the filenames: the image file is denoted as image_000,X, where X is the number of the image in the directory. But each directory has a different number of files, so we should use the list.files() function to collect the names of all the image files within the directories. We use them in the following section of code. The contents of the guitar photos directory when using the list.files() function are shown in a truncated form in Figure 11-9.

```
[1] "image_0001.jpg" "image_0002.jpg" "image_0003.jpg"
[9] "image_0009.jpg" "image_0010.jpg" "image_0011.jpg"
```

Figure 11-9. List of files from image directory

CHAPTER 11 ■ DEEP LEARNING AND OTHER EXAMPLE PROBLEMS

Now that we have the names of the individual files, we can load them into the img_data data frame using the following process:

```
#Creating Empty Data Frame
img_data <- data.frame()

#Turning Photos into Bitmaps
#Guitar Bitmaps
for (i in 1:length(bass_photos)){
  img <- readJPEG(paste("/path/to/image/directory/", guitar_photos[i], sep = ""))
```

We use the paste function here to combine the directory with the image with the string such that it leads us to the data. Using the readJPEG() function from the jpeg package, we can read the image into a bitmap, as described earlier as the stack of matrices. Each dimension represents the three colors (red, blue, and green) that make up every color photo. But to reduce the complexity of the images we're working with, we're going to convert these images to greyscale (black and white). When working with black and white images, we assign the pixel values a number between 0 and 1, with 0 representing black and 1 representing white. The colors in between determine the degree of intensity toward either side of the spectrum a particular color:

```
#Reshape to 64x64 pixel size and grayscale image
img <- Image(img, dim = c(64, 64), color = "grayscale")

#Resizing Image to 28x28 Pixel Size
img <- resize(img, w = 28, h = 28)
img <- img@.Data
```

We perform the reshaping and resizing of various images using the resize() function provided in EBImage. If you're interested in viewing what images look like when they're grayscaled, feel free to experiment with the display() and Image() functions accordingly. After the image is resized, we take the bitmap and convert it into a vector for a better storage method. Finally, we must add a label to the vector of data for when we're creating and training a model. This will be useful when calculating the accuracy of our model. Specifically, guitars will be labeled as 1 and laptops will be labeled as 2:

```
  #Transforming to vector
  img <- as.vector(t(img))

  #Adding Label
  label <- 1

  img <- c(label, img)

  #Appending to List
  img_data <- rbind(img_data, img)

}
```

We repeat this process for the laptop images. If you want to use this structure of preprocessing and model evaluation, feel free to do so—or experiment with alternative preprocessing methods. Prior to creating the CNN model, we must ensure that the input format for the model is correct. MXNet and many neural network models have specific formats that you should be familiar with. The first step is to create a training and test set. For this example, we'll be splitting the data set such that we train against 75% of the data and test against the remaining 25%. We now will transform the data such that it was a matrix in which each row was a different image observation, with the label as the first column entry and the bitmap values as the successive column entries. We'll then strip the label from the X matrix and use this as the values in the corresponding order of observations for the y vector. We then perform cross-validation using the `sample()` function:

```
#Transforming data into matrix for input into CNN
training_set <- data.matrix(img_data)

#Cross Validating Results
rows <- sample(1:nrow(training_set), nrow(training_set)*.75)

#Training Set
x_train <- t(training_set[rows, -1])
y_train <- training_set[rows, 1]
dim(x_train) <- c(28,28, 1, ncol(x_train))
```

In the preceding code, it's important to point out a distinct detail that if omitted will prevent you from being able to execute your code. The MXNet CNN model *only* takes an X matrix that is 4 dimensions. Be *sure* to remember this—otherwise you'll waste time debugging this issue! We also alter the dimensions of the test set accordingly:

```
#Test Set
x_test <- t(training_set[-rows, -1])
y_test <- training_set[-rows, 1];
dim(x_test) <- c(28,28, 1, ncol(x_test))
```

Now that we've finished preprocessing our data, we can finally begin to build and train our model.

Model Building and Training

CNN models are built in such a way that the data passes through each layer, but the only layer that's actually inputted to the `FeedForward()` function is the final layer. So we build the model prior to it being activated here. Some packages might be more proprietary and require less architecture, but MXNet allows for a significant degree of customization that would be useful if you would like to construct different ConvNet structures, such as those elaborated upon in Chapter 5. If you would like to improve upon the results here, that may be a good use of your time.

Let's move to the architecture. We'll be using a generic LeNet architecture here, as is the standard for image recognition tasks. As such, we organize the layers in the same manner:

```
data <- mx.symbol.Variable('data')

#Layer 1
convolution_l1 <- mx.symbol.Convolution(data = data, kernel = c(5,5),
num_filter = 20)
tanh_l1 <- mx.symbol.Activation(data = convolution_l1, act_type = "tanh")
pooling_l1 <- mx.symbol.Pooling(data = tanh_l1, pool_type = "max", kernel =
c(2,2), stride = c(2,2))

#Layer 2
convolution_l2 <- mx.symbol.Convolution(data = pooling_l1, kernel = c(5,5),
num_filter = 20)
tanh_l2 <- mx.symbol.Activation(data = convolution_l2, act_type = "tanh")
pooling_l2 <- mx.symbol.Pooling(data = tanh_l2, pool_type = "max",
kernel = c(2,2), stride = c(2,2))
```

We first start by creating a dummy data variable that will be used to pass the x matrix values in a file format friendly to the ConvNet here. data passes through each layer, as discussed in Chapter 5, where the model builds from lower abstractions to higher abstractions of the data to make a determination. Here, we will use a stride of 2 as generally recommended, 20 filters in the first Conv layer, and 50 filters in the second Conv layer. As an activation function, we use tanh. This activation function will be held constant throughout the entire model with the exception of the output function:

```
#Fully Connected 1
fl <- mx.symbol.Flatten(data = pooling_l2)
full_conn1 <- mx.symbol.FullyConnected(data = fl, num_hidden = 500)
tanh_l3 <- mx.symbol.Activation(data = full_conn1, act_type = "tanh")

#Fully Connected 2
full_conn2 <- mx.symbol.FullyConnected(data = tanh_l3, num_hidden = 40)

#Softmax Classification Layer
CNN <- mx.symbol.SoftmaxOutput(data = full_conn2)
```

The data continues to pass to the fully connected layers. Respectively, there are 500 and 40 hidden neurons in the fully connected layers. Finally, the data reaches the last layer, where we have a softmax classifier to determine the class of the observations.

CHAPTER 11 ■ DEEP LEARNING AND OTHER EXAMPLE PROBLEMS

Before we make any predictions, though, we must train our parameters using the method suggested in the previous section. When possible, particularly in the case of neural networks, using a local search method for packages that support these functionalities is highly recommended. Specifically, h2o supports a grid search function to tune parameters. Although here we're using MXNet, it's useful for readers to be aware of packages that do provide these functionalities.

Let's begin by training the parameters:

```
#Learning Rate Parameter
AUC <- c()
learn_rate <- c(0.01, 0.02, 0.03, 0.04)
CPU <- mx.cpu()

for (i in 1:length(learn_rate)){
  cnn_model <- mx.model.FeedForward.create(CNN, X = x_train,
  y = y_train, ctx = CPU, num.round = 50, array.batch.size = 40,
learning.rate = learn_rate[i],
momentum = 0.9, eval.metric = mx.metric.accuracy,
epoch.end.callback = mx.callback.log.train.metric(100),
 optimizer = "sgd")
#Code redated partially, please check github!
```

Similar to other neural network models, the learning rate parameter determines the magnitude of the gradient in updating the weights connecting the layers to each other. We give an array and plot the AUC, with respect to the tuning parameter in Figure 11-10.

CHAPTER 11 ■ DEEP LEARNING AND OTHER EXAMPLE PROBLEMS

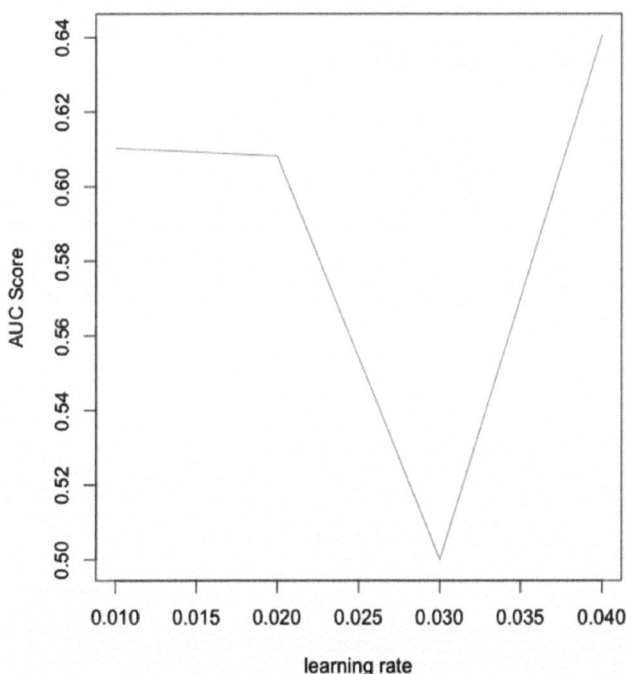

Figure 11-10. AUC score over learning rate

We can clearly see that a learning rate of 0.04 here is the most optimal because it yields the highest AUC score.

Let's now train the momentum parameter:

```
AUC1 <- c()
mom <- c(0.5, 0.9, 1.5)
for (i in 1:length(mom)){
cnn_model <- mx.model.FeedForward.create(CNN, X = x_train, y = y_train, ctx
= CPU, num.round = 50, array.batch.size = 40, learning.rate = 0.04,
momentum = mom[i], eval.metric = mx.metric.accuracy,
epoch.end.callback = mx.callback.log.train.metric(100), optimizer = "sgd")
#Code redacted partially, please check github!
```

CHAPTER 11 ■ DEEP LEARNING AND OTHER EXAMPLE PROBLEMS

When we execute the preceding code, we receive the results shown in Figure 11-11.

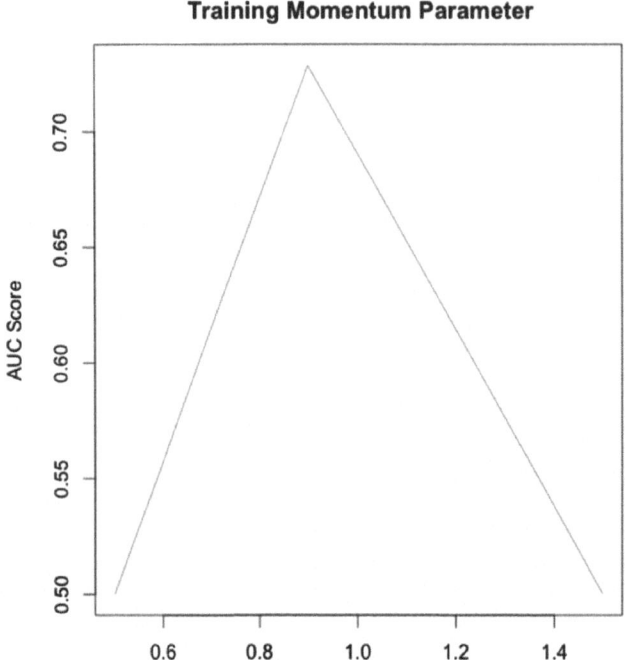

Figure 11-11. *AUC over momentum value*

When evaluating the results from the different parameters, we shall set the momentum value as 0.9. Now that we've tuned these two parameters, we can start training the tuned model in the final section and evaluating its performance on the test and training set:

```
#Fitted Model Training
cnn_model <- mx.model.FeedForward.create(CNN, X = x_train, y = y_train, ctx
= CPU, num.round = 150, array.batch.size = 40,
learning.rate = 0.04, momentum = 0.9, eval.metric = mx.metric.accuracy,
initializer = mx.init.normal(0.01) , optimizer = "sgd")

#Calculating Training Set Accuracy
y_h <- predict(cnn_model, x_train)
Labels <- max.col(t(y_h)) - 1
roc(as.factor(y_train), as.numeric(Labels))
curve <- roc(as.factor(y_train), as.numeric(Labels))
#Code partially redacted, please check github!
```

CHAPTER 11 ■ DEEP LEARNING AND OTHER EXAMPLE PROBLEMS

Before executing the code, I would like to point out one detail. Here, we have not enabled GPU training. If you want to decrease training time and improve computational performance, look into the necessary steps in the MXNet documentation to enable this feature. For this example, we'll be using CPU training. You should also be aware that the temptation to increase the num.round parameter will often be strong, as this will directly affect the accuracy of the model on the training set data. Beware that setting this parameter too high will cause overfitting, particularly on a data set the size of the one we're using in this example. When executing the preceding code, the user should see the terminal printing out the training accuracy in a format such as the following:

```
[184] Train-accuracy=0.708333333333333
[185] Train-accuracy=0.708333333333333
[186] Train-accuracy=0.708333333333333
[187] Train-accuracy=0.708333333333333
[188] Train-accuracy=0.708333333333333
```

The number on the left side of the words Train-accuracy represents the current iteration, which will run until the number indicated in the num.round parameter. The accuracy parameter used here is equivalent to the AUC score and is given by the mx.metric.accuracy object. As always, learning rates are difficult to approximate, but we can mitigate the loss of accuracy by adjusting the weights within the neural network using the stochastic gradient descent optimizer. When executing the code, we yield Figure 11-12.

CHAPTER 11 ■ DEEP LEARNING AND OTHER EXAMPLE PROBLEMS

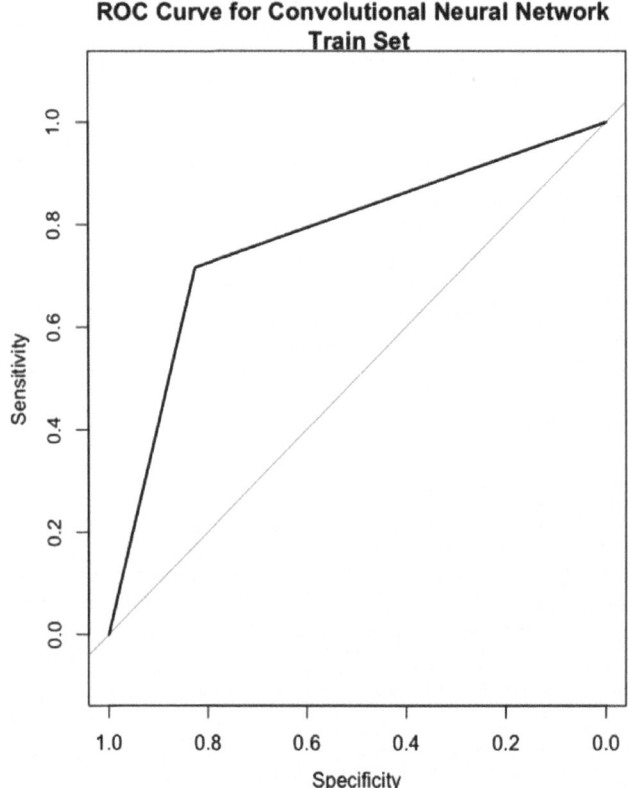

Figure 11-12. *ROC plot for CNN over training data*

This ROC plot has an AUC of 0.7706. When assessing the performance on the test data, Figure 11-13 and results are yielded.

CHAPTER 11 ■ DEEP LEARNING AND OTHER EXAMPLE PROBLEMS

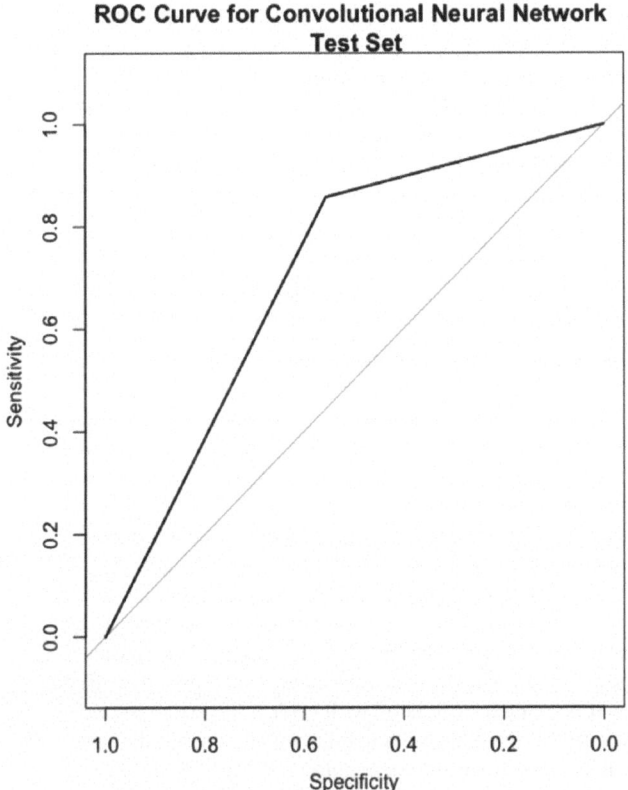

Figure 11-13. *ROC plot for CNN over test data*

The model when predicting against the test data has an AUC of 0.7063. Roughly speaking, the performance here is considerably similar, although as expected we do notice a drop-off in performance from the training to test set. That said, it's unlikely even in this instance that there is any indication of overfitting. However, you might still be inclined to improve the performance of these models should you want to put something like this into production. Ideally, when classifying images, we would like to have models that perform with at least 90% accuracy. Although the image classification case here is rather benign, there are cases in which incorrect classifications can lose considerable amounts of money per observation or cause incorrect diagnoses that therefore cause patients to receive improper care. With this in mind, how would you proceed from this point?

The most logical next step would be to acquire more data for the training phase of this model. This is typically what we would consider the largest challenge of building sufficient convolutional neural networks: getting enough training data. For many different commercial products, acquisition of this data legally can prove an exhaustive task and in a worst-case scenario would require acquisition of the data by the team itself in the real world. Readers should be mindful of this when creating CNNs for specific tasks, because sometimes the feasibility of the task is purely a matter of how accessible the data is. In this instance, the data set we're using is roughly only in total 170 photos, of which we train over 75%.

CHAPTER 11   DEEP LEARNING AND OTHER EXAMPLE PROBLEMS

Another suggestion you might want to take note of is using another network architecture, or if you feel ambitious enough, trying to create your own. However, creating your own network architecture can be an extremely daunting task. Another possible avenue to explore is creating several convolutional neural networks. From these models, we can create a data set where each feature is the output of a given CNN. This data set could then be inputted into a traditional machine learning model. You should be conscious, though, that these approaches might in an of themselves require significant tuning of the approach outlined earlier.

Collaborative Filtering

For our final example, we'll be briefly tackling the problem of recommendation systems as was briefly addressed in prior chapters. Recommendation systems are constantly evolving, but it's useful to address the concept because of the application of data science within them. It's here that you'll be introduced to the practical applications of imputation in addition to some of the soft skills of data science such as data transformation that have been briefly addressed but never walked through.

Recommendation systems are particular to e-commerce websites like Amazon. com but are also present in content-based sites such as Netflix. The motivation is fairly straightforward in that it is reasonable to recommend products to customers that they would reasonably like. The task of doing so is more difficult than it seems, though. Most users don't use the entirety of all products offered by a given company. Even if they did, it doesn't mean they would rate every single product they used. That leaves us with the problem of having a matrix that is sparsely populated with values. Nevertheless, we've reviewed techniques to handle this and will be moving on to inspecting our data set.

For this experiment, we'll be using the third Jester data set (http://goldberg. berkeley.edu/jester-data/). The features all represent individual jokes, and the rows represent users. Each entry within the matrix is a rating for a joke, where the lower bound is –10 and the upper bound is 10. However, whenever there isn't an entry for a joke, this is represented by a 99. When inspecting the head of the data set, we see the matrix shown in Figure 11-14.

```
     2      3   4   5     6   7     8     9  10
1   99  99.00  99  99  -9.27  99  -9.17  -8.59  99
2   99  99.00  99  99  -6.12  99  -7.48  -7.77  99
3   99   0.05  99  99  -2.82  99  -4.85  -0.87  99
4   99  99.00  99  99  -4.95  99   6.21   2.72  99
5   99  99.00  99  99   3.11  99   4.42   1.41  99
6   99  99.00  99  99  -0.05  99  -8.11  -7.38  99
```

Figure 11-14. *Snapshot of the Jester data set*

CHAPTER 11 ■ DEEP LEARNING AND OTHER EXAMPLE PROBLEMS

The goal here will be to measure the similarity of different users' tastes based on the similarities between the jokes themselves. To do this, we'll be calculating the cosine similarity between column vectors. Briefly, let's discuss the concept of cosine similarity before speaking about combining matrix factorization and RBMs to impute missing values. When working with problems in which you're trying to compare vectors, cosine similarity is a concept that will often be referenced. Intuitively, we define *cosine similarity* as the degree to which two non-zero vectors are distinct. Mathematically, we define cosine similarity with the following equation

$$\text{similarity} = \cos(\theta) = \frac{A * B}{\|A\|_2 \|B\|_2}$$

where A, B = two distinct vectors.

Similarly to a correlation coefficient, cosine similarity values range from –1 to 1. A cosine similarity of 1 indicates that values are exactly the same, whereas –1 means they are exactly opposite. A value of zero indicates no relationship between vectors at all. With this in mind, we'll compare the consumption patterns of certain music with one another such that we can compare which items are most like each other and therefore should be recommended to other individuals.

However, for those who paid close attention, cosine similarity is used with two *non-zero* vectors—meaning we have to generate values for our dataset where they are missing. There are many techniques that have been discussed for imputation, but one that has been described as useful by Geoffrey Hinton in this instance is *matrix factorization*. Specifically, I suggest you use *singular value decomposition* (SVD).

SVD and PCA, discussed elsewhere in this book, are highly related techniques. They both are perform eigendecompositions of a matrix, but SVDs applications differ from that of PCA. Particularly, SVD can be used to approximate the missing values. As such, let's impute our values using the impute.svd() function:

```
require(lsa)
require(bcv)
require(gdata)
require(Matrix)

#Upload the data set
#Please be patient this may take a handful of seconds to load.
data <- read.xls("/path/to/data/.xls", sheet = 1)
colnames(data) <- seq(1, ncol(data), 1)

#Converting 99s to NA Values (1)
data[data == 99] <- NA

#Converting 99s to Mean Column Values (2)
for (i in 1:ncol(data)){
  data[is.na(data[,i]), i] <- mean(data[,i], na.rm = TRUE)
}
```

We begin by converting the 99s (1) to NA values and then changing the NA values to the column means (2). After this point, we can move forward and impute the values:

```
#Imputing Data via SVD
newData <- impute.svd(data, k = qr(data)$rank, tol = 1e-4, maxiter = 200)
print(newData$rss)
head(data[, 2:10])
```

Be aware that the impute.svd() function requires that you impute either column means for the missing values, or if an entire column's observations are missing to make it 0. If you don't follow these instructions, you'll receive incorrect results. When executing the preceding code, we yield the outputs shown in Figure 11-15.

2	3	4	5	6	7	8	9	10
1.987538	1.336403	1.506	1.83	-9.27	4.056429	-9.17	-8.59	-1.8
1.987538	1.336403	1.506	1.83	-6.12	4.056429	-7.48	-7.77	-1.8
1.987538	0.050000	1.506	1.83	-2.82	4.056429	-4.85	-0.87	-1.8
1.987538	1.336403	1.506	1.83	-4.95	4.056429	6.21	2.72	-1.8
1.987538	1.336403	1.506	1.83	3.11	4.056429	4.42	1.41	-1.8
1.987538	1.336403	1.506	1.83	-0.05	4.056429	-8.11	-7.38	-1.8

Figure 11-15. *Head of imputed data set*

When executing the SVD, we also calculated a sum of squares of 4.398197e-20 with respect to the non-missing values and the predictions of these non-missing values. Readers who feel inclined to challenge themselves here can, instead of using SVD impute the values, use an RBM. Be aware, though, that this task can be extremely computationally expensive, and the modification of the RBM for this task is not easy. Look for high-level overviews given by Geoffrey Hinton on this topic (http://www.machinelearning.org/proceedings/icml2007/papers/407.pdf).

We can now calculate the cosine distances between the columns:

```
itemData <- matrix(NA, nrow = ncol(data), ncol = 11,
                   dimnames=list(colnames(data)))
#Getting Cosine Distances
for (i in 1:nrow(itemData)){
  for (j in 1:ncol(itemData)){
    itemData[i,j] <- cosine(data[,i], data[,j])
  }
}
```

CHAPTER 11 ■ DEEP LEARNING AND OTHER EXAMPLE PROBLEMS

When executing the preceding code, we yield the data set shown in Figure 11-16.

```
    [,1]       [,2]       [,3]       [,4]       [,5]       [,6]       [,7]       [,8]
1 1.00000000 0.96586519 0.93061731 0.97344905 0.97438177 0.03342678 0.97393042 -0.1563475
2 0.96586519 1.00000000 0.95495652 0.99103253 0.99129663 0.03424918 0.99084916 -0.1648979
3 0.93061731 0.95495652 1.00000000 0.95527979 0.95531230 0.03379699 0.95469894 -0.1585386
4 0.97344905 0.99103253 0.95527979 1.00000000 0.99928582 0.03380817 0.99849612 -0.1672893
5 0.97438177 0.99129663 0.95531230 0.99928582 1.00000000 0.03365341 0.99952555 -0.1675603
6 0.03342678 0.03424918 0.03379699 0.03380817 0.03365341 1.00000000 0.03385248  0.2222439
```

Figure 11-16. *Head of the cosine distance data set*

From this data set, we can now perform the final data transformation such that each row represents a particular joke and each column represents the jokes most similar in descending order from left to right. We do this initially by instantiating an empty matrix with the proper dimensions (1). After this matrix is instantiated, we can then fill in the data by sorting the cosine values and taking the indices that contain the top 11 values—we take the top 11 because the number 1 value will always be the same item itself:

```
#Creating Matrix for ranking similarities (1)
similarMat <- matrix(NA, nrow = ncol(itemData), ncol = 11)

#Sorting Data Within Item Data Matrix (2)
for(i in 1:ncol(itemData)) {
  rows <- order(itemData[,i], decreasing = TRUE)
  similarMat[i,] <- (t(head(n=11, rownames(data[rows ,][i]))))
}

#Printing Result
similarMat
```

When executing the preceding code, we reach our final answer, shown in Figure 11-17.

```
       [,1] [,2] [,3] [,4] [,5] [,6] [,7] [,8]  [,9] [,10] [,11]
 [1,]  "1"  "5"  "7"  "4"  "26" "31" "11" "2"   "101" "12"  "15"
 [2,]  "2"  "5"  "4"  "7"  "26" "11" "31" "101" "12"  "15"  "13"
 [3,]  "3"  "5"  "4"  "2"  "7"  "11" "26" "31"  "101" "12"  "15"
 [4,]  "4"  "5"  "7"  "26" "31" "11" "101" "12" "15"  "13"
 [5,]  "5"  "7"  "4"  "26" "31" "11" "2"  "101" "12"  "15"  "13"
 [6,]  "6"  "20" "21" "28" "8"  "18" "37" "19"  "36"  "51"  "54"
 [7,]  "7"  "5"  "4"  "26" "31" "11" "2"  "101" "12"  "15"  "13"
 [8,]  "8"  "21" "19" "16" "20" "18" "9"  "14"  "17"  "6"   "75"
 [9,]  "9"  "16" "14" "18" "17" "10" "25" "24"  "59"  "65"  "58"
[10,]  "10" "24" "25" "59" "58" "75" "72" "45"  "38"  "34"  "61"
[11,]  "11" "5"  "7"  "4"  "26" "2"  "31" "12"  "101" "15"  "13"
```

Figure 11-17. *Top 10 recommendations for 11 separate jokes*

We interpret the result as yielding the top 10 recommendations for 11 separate jokes. You can implement this in a platform such that on a web page, users receive recommendations for different pages, products, and or similar entities.

Summary

We now have reached the end of this chapter and our review of deep learning and machine learning techniques entirely. Chapter 12 provides brief advice that all data scientists should be aware of as they move forward in their research or professional endeavors.

CHAPTER 12

Closing Statements

We have reached the end of this book. By now, you should feel comfortable that you've acquired a general overview of data science, machine learning, and deep learning. If not, you should at least be adequately aware of where you need to focus your efforts in reviewing and further research. The purpose of this book is *not* intended to make anyone an expert. Rather it should be used to highlight the respective power of these techniques in a given field. I would like to end by imparting advice for all readers with my thoughts on the best way to use these models and the general methodology of machine learning.

In every field, there are idiosyncratic characteristics that have long been studied. This is generally what I would describe as the *science of X*, where X is a given field we're discussing. Sometimes specific quantitative subfields have been developed within the broader field to tackle these goals. Given the complex nature of the world, it can't be overemphasized that studying the broader field and the specific subfield you're interested holistically is a requirement *before* you seek to implement machine learning methods to problem solving. Among the complaints I have heard from many colleagues and friends is that there is one overwhelming deficiency that many data scientists often have: *domain knowledge*. Machine learning and deep learning algorithms have gotten very good at performing in a variety of contexts and increasingly have been able to produce robust solutions. however, using a good tool poorly in a given context can produce results just as bad, if not worse, as using the wrong tool poorly in a given context.

You should be sure that you deeply understand the algorithms you choose prior to implementing them at scale. There is seldom anything worse than providing a solution, seeing the process for it fail, and being unable to provide counsel on how to fix it. Beyond something bad happening, often you'll be expected to discuss these algorithms with people who have less technical backgrounds. Although I have emphasized this in prior chapters, I must again state the power of good visualizations and succinct explanations. Although you might find intricate detail compelling, the average person doesn't have the time that you have spent educating yourself on this topic. So, only make things as complicated as they need to be.

Finally, I urge you when producing your own solutions to be as creative as possible. The proliferation of machine learning algorithms is exciting for the way in which it revolutionizes our world, but it will also lead to great homogeneity among products if these algorithms aren't used in a unique way. The process of solving problems,

CHAPTER 12 ■ CLOSING STATEMENTS

while frustrating at times, should be challenging and exciting also. This should be an opportunity to use your ingenuity to create a unique solution—not use old and tired solutions. Reusing large portions of code, though tempting and often necessary to save time in the development stage, should be avoided as much as possible as well. Always force yourself to approach a problem from scratch, because that will inspire new and hopefully better solutions.

I wish all readers the best of success moving forward in their respective studies and careers, and also in life. Machine learning is one of the most frustrating concepts I have ever encountered, but through studying it I've learned an incredible amount about computer science and myself while also being introduced to an immense amount of incredibly intelligent individuals. I hope that the joy that has been brought into my life from studying this field is similarly brought to yours Godspeed.

Sincerely,
Taweh Beysolow II

Index

A

A/B testing, 148
    beta-binomial hierarchical model for, 149, 151
    simple two-sample, 149
Activation function, 3
Additive law of probability, 13
Akaike information criterion (AIC), 152
AlexNet, 110
Amazon Web Services (AWS), 167
Analysis of Variance (ANOVA), 137
    MANOVA, 138
    mixed-design, 138
    one-way, 137
    two-way (multiple-way), 137
Ant colony optimization (ACO), 159–160
Arithmetic mean, 15
Asset price prediction, 171–172
    description of experiment, 173–175
    feature selection, 175–176
    supervised learning, 172–173
Associative property, 19
Autoencoders, 125–126, 195–199, 201–202
    linear autoencoders *vs.* PCA, 126–127
Axioms, 19
    associative property, 19
    commutative property, 19
    distributivity of scalar multiplication, 20
    identity element of addition, 19
    identity element of scalar multiplication, 20
    inverse elements of addition, 19

B

Back-propagation algorithm, 95–97, 107
Back-propagation through time (BPPT), 114–115
Backward selection, 151
Batch learning, 131
Bayesian classifier, 191–193, 196
Bayesian learning, 83
    50/25/25 cross-validation, 85–86
    limitation, 84
    Naïve Bayes classifier, 84
    tuning machine learning algorithms, 85
Bayesian statistics, 149
Bayes information criterion (BIC), 152
Bayes' theorem, 14
Beta-binomial hierarchical model, 149, 151
Beta distribution, 150
Bi-infinite sequence, 40
Binary classifier, 66
Binomial distribution, 150, 151
Blocking process, 145
BPTT. *See* Back-propagation through time (BPTT)

C

Canonical correlation analysis (CCA), 156
Central processing unit (CPU), 168, 169
Coefficient of determination (R squared), 17
Collaborative filtering, 214–218
Commutative property, 19
Complex cells, 101
Confusion matrix, 68–69
    for Bayesian classifier, 85
    for classification tree, 80

■ INDEX

Conjugate distribution, 150
Conjugate gradient algorithms, 98, 119
Continuous random variables, 14
Contrasting divergence (CD)
 learning, 129-131
Convergent sequence, 41
Convolutional layer
 convolving, 104
 feature maps, 104
 filtering, 104
Convolutional neural networks
 (CNNs), 5, 108, 202, 204
 AlexNet, 110
 convolutional layer, 103-104
 depth, 108
 FC layer, 106
 GoogLeNet, 109
 history, 101
 loss layer, 107
 pooling layer, 105
 preprocessing, 204-206
 regularization, 111
 ReLU layer, 106
 ResNet, 110
 structure and properties, 101-103
 tuning parameters, 108
 VGGnet, 110
Convolving, 104
Cook's distance, 142
Correlation coefficients, 16
Correlation matrix, 175
Cosine similarity, 215

■ D

Data science, 219
Decision tree learning, 78
 classification trees, 79-80
 limitations, 81
 regression trees, 80-81
Deep belief network (DBN), 6, 134-135
Deep learning, 219
 autoencoders, 195-199, 201-202
 CNNs, 202, 204
  preprocessing, 204-206
 model building and training, 206-214
  collaborative filtering, 214-218
 models, 3
  applied machine learning and, 7
  CNNs, 5
  DBNs, 6

experimental design, 7
feature selection, 7
history, 8
MLP, 4
restricted Boltzmann machines, 6
RNNs, 5
SLP, 3-4
structure of, 2
Deep neural networks, 2
Derivatives and differentiability, 42
Diagonal matrix, 22
Discrete random variables, 14
Discriminant, 43
Distributional prediction, 79
DropConnect, 111
DropOut, 111

■ E

Eigenvalues, 34-36
Eigenvectors, 34-36
Elman neural networks, 115
Embedded algorithms, 157. *See also*
 Wrappers, Filters, and
 Embedded (WFE) algorithms
Ensemble methods, 82
 gradient boosting algorithm, 82-83
 random forest, 83
  limitations, 83
Euclidean function
 Euclidean loss, 107
 softmax loss function, 107
 softmax normalization, 107
Euclidean norm, 29
Expectation maximization (EM)
 algorithm, 76
 expectation step, 77
 maximization step, 77-78

■ F

Factor analysis, 154-155
 limitations, 155
Factor loadings, 155
Fast learning algorithm, 135
 steps, 136
Feature maps, 104
Feature/variable selection
 techniques, 151
 backwards and forward
  selection, 151-152

factor analysis, 154–155
    limitations, 155
    PCA, 152–154
FeedForward() function, 206
Fisher's principles, 144–145
Fixed tabu search, 163–164
F-statistic and F-distribution, 138–145
Full factorial, 147
Fully connected (FC) layer, 102, 106
Fully recurrent networks, 113–114

G

Genetic algorithms (GAs), 158
Geometric mean, 15
Gibbs sampling, 129, 135
Global minimizers, 47–48
Global optimum, 95
Google Finance API, 172
GoogLeNet, 109
Gradient, 42
Gradient boosting algorithm, 82–83
Gradient descent algorithm, 53–54
Graphics processing unit (GPU), 168

H

Hadamard matrix, 146
Halton, Faure, and Sobol sequences, 148
Hamming distance, 164
Handling categorical data, 155
    categorical label problems, 156
    CCA, 156
    encoding factor levels, 156
Hard drive disk (HDD), 167
Hardware and software suggestions
    CPU, 169
    GPU, 168
    motherboard, 169
    optimizing machine learning software, 170
    processing data with standard hardware, 167
    PSU, 170
    RAM, 169
    solid state drive and HDD, 167
Having memory, 113, 116
Hessian-free optimization, 48
Hessian matrix, 43, 49
Hidden layers, 99
Hill climbing search methods, 158

I

Identity matrix, 22
ImageNet Large-Scale Visual Recognition Challenge (ILSVRC), 109–110
Inception architecture, 109
Instantaneous algorithm, 90
Intelligent optimization, 161
Interpretation, 145

J

Jacobian matrix, 49

K

Kernels, 72
K-means clustering, 74, 156
    limitations, 75–76
K-nearest neighbors (KNN), 165, 189–191

L

Learning rate, 54–55, 209
    choosing, 55–56, 58–60
    Levenberg-Marquardt heuristic, 61
    Newton's method, 60–61
Least Absolute Shrinkage and Selection Operator (LASSO), 63
    ridge regression and, 64
LeNet, 108
Levenberg-Marquardt (LM) algorithm, 61
Leverage, 142
Linear algebra, 17
    axioms, 19
        associative property, 19
        commutative property, 19
        distributivity of scalar multiplication, 20
        identity element of addition, 19
        identity element of scalar multiplication, 20
        inverse elements of addition, 19
    matrices, 20
        addition, 21
        column vector and square matrix, 24
        derivatives and differentiability, 42
        distributive over matrix addition, 27

Linear algebra(*cont.*)
    eigenvalues and
        eigenvectors, 34–36
    Euclidean norm, 29
    Hessian, 43
    hyperplanes, 39–40
    inner products, 32
    L1 norm, 29–30
    L2 norm, 29
    limits, 41
    linear transformations, 36–37
    matrix by matrix
        multiplication, 23
    multiplication, 22
    multiplication properties, 26
    norms, 29, 31
    norms on inner product
        spaces, 32–33
    nullspace, 39
    orthogonality, 34
    orthogonal projections, 38
    outer product, 34
    partial derivatives and
        gradients, 42
    P-norm, 30–31
    proofs, 33–34
    properties, 21
    quadratic forms, 37
    range, 38
    rectangular, 26
    row and column vector
        multiplication, 24
    row vector, square matrix, and
        column vector, 25
    scalar multiplication, 21, 23, 27
    sequences, 40
    sequences, properties, 40
    square, 25
    Sylvester's criterion, 37–38
    trace, 28
    transpose, 28
    transposition, 21
    types, 21–22
    scalars and vectors, 17
    subspaces, 20
    vectors, properties, 18
        addition, 18
        element wise multiplication, 19
        subtraction, 18
Linear autoencoders *vs.* PCA, 126–127
Linear regression, 51
    gradient descent algorithm, 53–54
    learning rate, 54–55
    multiple linear regression via gradient
        descent, 54
    OLS, 51–53
Linear transformation, 36–37
Local minimizers, 47
    conditions for, 48–49
Local search methods, 157
    ACO, 159–160
    genetic algorithms (GAs), 158
    hill climbing, 158
    simulated annealing (SA), 159
    VNS, 160–161
Logistic function, 66
Logistic regression, 66–67, 186–189
    limitations, 69–70
Long short-term memory (LSTM)
    applications, 117
    distinguishing factor, 117
    forget gate, 117
    overview, 116
    traditional, 118
    training, 118
    visualization, 117
Loss layer, 107

M

Machine learning, 1, 50, 219
    algorithms, 51
    asset price prediction, 171–172
        description of experiment, 173–175
        feature selection, 175–176
        supervised learning, 172–173
    feature selection, 185–186
    history, 50
    model evaluation, 176
        ridge regression, 176–178
        SVR, 178–180
    model training and evaluation, 186
        Bayesian classifier, 191–193
        KNN, 189–191
        logistic regression, 186–189
    proliferation, 219
    speed dating, 180
        classification, 181–182
        data cleaning and
            imputation, 182–185
    unsupervised learning, 74
        assignment step, 74

K-means clustering, 74
K-means clustering,
   limitations, 75–76
   update step, 75
Markov process, 87
Matrices, 20
   addition, 21
   column vector and square matrix, 24
   derivatives and differentiability, 42
   distributive over matrix addition, 27
   eigenvalues and eigenvectors, 34–36
   Euclidean norm, 29
   Hessian, 43
   hyperplanes, 39–40
   inner products, 32
   L1 norm, 29–30
   L2 norm, 29
   limits, 41
   linear transformations, 36–37
   matrix by matrix multiplication, 23
   multiplication, 22
   multiplication properties, 26
   norms, 29, 31
   norms on inner product spaces, 32–33
   nullspace, 39
   orthogonality, 34
   orthogonal projections, 38
   outer product, 34
   partial derivatives and gradients, 42
   P-norm, 30–31
   proofs, 33–34
   properties, 21
   quadratic forms, 37
   range, 38
   rectangular, 26
   row and column vector
      multiplication, 24
   row vector, square matrix, and column
      vector, 25
   scalar multiplication, 21, 23, 27
   sequences, 40
      properties, 40
   square, 25
   Sylvester's criterion, 37–38
   trace, 28
   transpose, 21, 28
   types, 21–22
Mean squared error (MSE), 17, 65
Mixed-design ANOVA, 138
mlp() function, 100

MLP. *See* Multilayer perceptron (MLP)
   model
Momentum within RBMs, 132
Motherboard, 169
Multicollinearity, 62
   confusion matrix, 68–69
   logistic regression, 69–70
   regression models, 64–67
   ridge regression, 62–64
   ROC curve, 67–68
   SVM, 70–73
   testing, 62
   VIF, 62
Multilayer perceptron (MLP) model, 4
   back-propagation algorithm, 95–97
   considerations, 97–99
   distinguishing factor from SLPs, 94
   global optimum, 95
   limitations, 97–99
Multiple linear regression via gradient
   descent, 54
Multiplicative law of probability, 13
Multivariate ANOVA (MANOVA), 138
Mxnet, 99

N

Naïve Bayes classifier, 84
Neighborhoods, concept, 49
   interior and boundary points, 50
Netflix, 214
Neural history compressor, 116
Newton's method, 60–61
Non-parametric bootstrapping, 81
Norms, 29
Null hypothesis, 145

O

One-way ANOVA, 137
Online learning, 131
Optimization, 45
   unconstrained, 45–46
      global minimizers, 47–48
      local minimizers, 47
      local minimizers,
         conditions, 48–49
Ordinary least squares (OLS), 51–53
Orthogonality, 34
Orthogonal projections, 38

P

Parameter tuning, 173
Partial derivative, 42
Perceptron model, training, 90
Plackett-Burman designs, 146
Point prediction, 79
Pooling layer, 105
Positive semi-definite matrix, 22
Posterior distribution, 149
Power supply unit (PSU), 170
Principal components, 152
Principal components analysis
    (PCA), 36, 126-127, 152-154, 176
Prior distribution, 149
Probability, 11-12
Probability theory, 86
Pseudo-random numbers, 148

Q

Quadratic forms, 37
Quantitative finance, 171

R

Random access memory (RAM), 169
Random forest, 83
    limitations, 83
Randomization, 145
Random sampling, 14
Random variables, 14-15
Reactive search optimization (RSO), 161
    fixed tabu search, 163-164
    KNN, 165
    reactive prohibitions, 162-163
    RTS, 164
    WalkSAT algorithm, 165
Reactive tabu search (RTS), 164
Receiver Operating Characteristic
    (ROC) curve, 67-68
Rectangular matrices, 26
Rectified linear units (ReLU) layer, 106
Recurrent neural networks (RNNs), 5
    architecture, 114
    BPPT, 114-115
    Elman, 115
    example, 120-124
    fully, 113-114
    LSTM, 116-118
    neural history compressor, 116

    parameter update algorithm, 119-120
    structural damping within, 119
Regression, 172-173
Regression models, 51
    evaluating, 64-65
        classification, 65
        coefficient of determination, 65
        logistic regression, 66-67
        MSE, 65
        SE, 65
    linear regression, 51
        gradient descent algorithm, 53-54
        learning rate, 54-55
        multiple linear regression via
            gradient descent, 54
        OLS, 51-53
Regularization
    DropConnect, 111
    DropOut, 111
    L1 and L2, 111
    negative effect, 111
    stochastic pooling, 111
Reinforcement learning, 86-87
Relief algorithm, 157
ResNet, 110
Restricted Boltzmann machines
    (RBMs), 6, 125, 127
    energy function, 127
    Hopfield networks, 128
    implementations, 129
    individual activation probabilities, 129
    momentum within, 132
    probability distributions, 128
    standard, 127
    visualization, 135
Ridge regression, 62, 63, 176-178
    and LASSO, 64
Robust tabu search, 163

S

Simple cells, 101
Simulated annealing (SA), 159
Single layer perceptron (SLP) model, 3-4
    activation function, 90
    architecture, 89
    distinguishing factor from MLP, 95
    limitations, 91-93
    perceptron model, 90
    statistics, 94
    WH algorithm, 90

Singular value decomposition (SVD), 215–216
SLP. *See* Single layer perceptron (SLP) model
Solid state drives, 168
Space filling, 147
Sparsity, 133
Speed dating, 180
    classification, 181–182
    data cleaning and imputation, 182–185
Standard deviation, 16
Standard error (SE), 65
Statement of experiment, 144
Statistical concepts, 11
    and *vs.* or, 12–13
    Bayes' theorem, 14
    coefficient of determination (R squared), 17
    MSE, 17
    probability, 11–12
    random variables, 14–15
    standard deviation, 16
    variance, 15
Statistical replication, 145
Stochastic pooling, 111
Stride, 108
Structural damping, 119
Subspaces, 20
Supervised learning, 50
    regression, 172–173
Support vector machine (SVM), 70–72
    extensions, 73
    kernels, 72
    limitations, 73
    sub-gradient method applied to, 72
Support vector regression (SVR), 178–180
Sylvester's criterion, 37–38

T

Test of significance, 145
Transposition, 18

Two-way infinite sequence, 40
Two-way (multiple-way) ANOVA, 137

U

Unconstrained optimization, 45–46
    global minimizers, 47–48
    local minimizer, conditions, 48–49
    local minimizers, 47
Unsupervised learning, 74
    assignment step, 74
    K-means clustering, 74
    limitations, 75–76
    update step, 75

V

Vanishing gradient, 116
Variable neighborhood search (VNS), 160–161
Variance, 15
Variance inflation factor (VIF), 62
VGGnet, 110

W, X

Wake-sleep algorithm, 8
WalkSAT algorithm, 165
Weight decay, 133
Widrow-Hoff (WH) algorithm, 90
Wrappers, Filters, and Embedded (WFE) algorithms, 157
    relief algorithm, 157

Y

Yahoo! Finance API, 172

Z

Zero-padding, 108

Get the eBook for only $5!

Why limit yourself?

With most of our titles available in both PDF and ePUB format, you can access your content wherever and however you wish—on your PC, phone, tablet, or reader.

Since you've purchased this print book, we are happy to offer you the eBook for just $5.

To learn more, go to http://www.apress.com/companion or contact support@apress.com.

Apress®

All Apress eBooks are subject to copyright. All rights are reserved by the Publisher, whether the whole or part of the material is concerned, specifically the rights of translation, reprinting, reuse of illustrations, recitation, broadcasting, reproduction on microfilms or in any other physical way, and transmission or information storage and retrieval, electronic adaptation, computer software, or by similar or dissimilar methodology now known or hereafter developed. Exempted from this legal reservation are brief excerpts in connection with reviews or scholarly analysis or material supplied specifically for the purpose of being entered and executed on a computer system, for exclusive use by the purchaser of the work. Duplication of this publication or parts thereof is permitted only under the provisions of the Copyright Law of the Publisher's location, in its current version, and permission for use must always be obtained from Springer. Permissions for use may be obtained through RightsLink at the Copyright Clearance Center. Violations are liable to prosecution under the respective Copyright Law.